Hydrology

Edited by Theodore V. Hromadka II
and Prasada Rao

Published in London, United Kingdom

IntechOpen

Supporting open minds since 2005

Hydrology
http://dx.doi.org/10.5772/intechopen.87673
Edited by Theodore V. Hromadka II and Prasada Rao

Contributors
Supriyo Chakraborty, Siddharth Birmal, Pramit Kumar Deb Burman, Amey Datye, Fousiya A. A., Aravind G. H., Mohan P. M., Neha Trivedi, R. K. Trivedi, Amin Shaban, Pawan Jeet, Alok Kumar, Prem Kumar Sundaram, Maciej Zalewski, Prasada Rao, Theodore V. V. Hromadka II, Miha Curk, Matjaž Glavan, Mohammad Valipour, Sayed Bateni, Nicolas R. Dalezios, Mansour Almazroui, Essam Heggy, Zekai Şen, Andreas N. Angelakis

First published in London, United Kingdom, 2021 by IntechOpen
IntechOpen is the global imprint of INTECHOPEN LIMITED, registered in England and Wales, registration number: 11086078, 5 Princes Gate Court, London, SW7 2QJ, United Kingdom
Printed in Croatia

British Library Cataloguing-in-Publication Data
A catalogue record for this book is available from the British Library

Additional hard and PDF copies can be obtained from orders@intechopen.com

Hydrology
Edited by Theodore V. Hromadka II and Prasada Rao
p. cm.
Print ISBN 978-1-83962-329-5
Online ISBN 978-1-83962-330-1
eBook (PDF) ISBN 978-1-83962-331-8

For EU product safety concerns:
IN TECH d.o.o., Prolaz Marije Krucifikse Kozulić 3, 51000 Rijeka, Croatia, info@intechopen.com or visit our website at intechopen.com.

Meet the editors

Hromadka & Associates' Principal and Founder, Theodore Hromadka II, Ph.D., Ph.D., Ph.D., PH, PE, has extensive scientific, engineering, expert witness, and litigation support experience. His frequently referenced scientific contributions to the hydrologic, earth, and atmospheric sciences have been widely published in peer-reviewed scientific literature, including 30 books and more than 500 scientific papers, book chapters, and government reports. His professional engineering experience includes supervision and development of over 1500 engineering studies. He is currently a faculty member at the United States Military Academy at West Point, New York.

Prasada Rao is a Professor in the Civil and Environmental Engineering Department at California State University, Fullerton. His current research areas relate to surface and subsurface flow modeling and computational mathematics. He has worked extensively on developing innovative hydraulic and hydrological modeling solutions to better predict surface flow phenomena along with their impact on groundwater levels. He has also worked on developing parallel hydraulic models for large scale applications. He has taught undergraduate and graduate level courses in hydraulics, hydrology, open channel flow, and hydraulic structures.

Contents

Preface

Recent advances in the interdisciplinary science of hydrology are facilitating a better understanding of the underlying physics associated with hydrologic cycle components, improved predictive capabilities of surface water numerical models, and enhanced strategic water management protocols. An effort is made in this book to present some of these advances. The book will be of interest to students who would like to advance their knowledge in hydrology and researchers in areas related to modeling, hydrometeorology, agronomy, water management policy decision-makers, environmental science, and earth sciences. The contents of this book will supplement the material in standard hydrology textbooks. A brief overview of the book's chapters is summarized below.

The science of hydrometeorology, which focuses on studying the atmospheric and terrestrial components of the hydrologic cycle with emphasis on their interrelations, is fast maturing. Advances made in instrumentation and computational power are enabling researchers to better understand the underlying science. Valipour and his team (Chapter 1) extensively reviewed the evolution of observational hydrometeorology from 3500 BC to the present day and their combined role with modeling in assessing climate change. Curk and Glavan (Chapter 2) reviewed the popular hydrologic models with a focus on their suitability and robustness for agronomy applications. The chosen model's strengths, weaknesses, and the importance of interpreting the model output were presented. Related studies in Slovenia have been summarized.

Legacy numerical models for hydrologic applications were originally written prior to 1990. These models were written in FORTRAN and are based on a strong theoretical foundation. However, owing to the lack of computational resources at that time, the coding techniques used are not optimal for the current computers. Additionally, the array limitations in the legacy codes might prevent their application over large domains. Rao and Hromadka (Chapter 3) reviewed the limitations in the USGS Diffusion Hydrodynamic Model (DHM) and enhanced certain components in the source code so that the model can be applied across large spatial domains. Zalewski (Chapter 4) detailed the underlying elements in ecohydrology, which is an interdisciplinary field that brings in hydrologists, hydrobiologists, botanists, and plant physiologists. The role of ecohydrology in achieving biosphere sustainability and the need to develop it as integrative sustainability science has been detailed.

Jeet and his colleagues, (Chapter 5) in their chapter, reviewed the effort that went in and the challenges associated with the linking of rivers in India. While linking of the rivers will address the water imbalance, addressing the associated infrastructure costs and potential opposition from landowners will require a significant commitment from the federal and state agencies. Some of the parameters that the authors have presented are very similar to other parts of the world, where discussion relating to interlinking rivers is in the infant stages. Shaban (Chapter 6) reviewed the characteristics of the rivers in Lebanon along with their watersheds.

The challenges associated with efficiently managing the nation's water resources coupled with the importance of developing an integrated management policy are detailed. Chakraborty and his colleagues (Chapter 7) detailed the statistical analysis of precipitation isotopes across gauge rainfall data from multiple stations in the Indian subcontinent. A potential application of this technique is in predicting the interannual variability of monsoon rainfall.

Theodore V. Hromadka II
Distinguished Professor,
Department of Mathematical Science,
United States Military Academy,
West Point, NY, USA

Prasada Rao
Professor,
Department of Civil and Environmental Engineering,
California State University,
Fullerton, CA, USA

Chapter 1

Hydrometeorology: Review of Past, Present and Future Observation Methods

Mohammad Valipour, Sayed M. Bateni, Nicolas R. Dalezios, Mansour Almazroui, Essam Heggy, Zekai Şen and Andreas N. Angelakis

Abstract

Hydrometeorology aims at measuring and understanding the physics, chemistry, energy and water fluxes of the atmosphere, and their coupling with the earth surface environmental parameters. Accurate hydrometeorological records and observations with different timelines are crucial to assess climate evolution and weather forecast. Historical records suggest that the first hydrometeorological observations date back to *ca* 3500 BC. Reviewing these observations in the light of our modern knowledge of the dynamic of atmospheres is critical as it can reduce the ambiguities associated to understanding major fluctuations or evolutions in the earth climate. Today, the ambiguities in hydrometeorological observations have significantly improved due to the advances in monitoring, modeling, and forecasting of processes related to the land-atmosphere coupling and forcing. Numerical models have been developed to forecast hydrometeorological phenomena in short-, medium- and long-term horizons, ranging from hourly to annual timescales. We provide herein a synthetic review of advances in hydrometeorological observations from their infancy to today. In particular, we discuss the role of hydrometeorological records, observations, and modeling in assessing the amplitude and time-scale for climate change and global warming.

Keywords: hydrometeorology, sustainability, weather monitoring tools, climate, forecasting, innovations

1. Introduction

In general, hydrometeorology deals with monitoring the energy and water fluxes between the atmosphere and earth [1–4]. Hydrometeorology has evolved as a special discipline of both meteorology and hydrology, linking the fundamental knowledge of meteorologists with the needs of hydrologists to assess the water and energy cycles at local, regional, and global scales [1–4]. In hydrometeorology, meteorological data are incorporated into hydrological models to predict water and energy exchanges between the land surface and atmosphere, weather, climate, and natural hazards such as wildland fires, storms, droughts, and floods [5–8]. Climatologists focus on seasonal to decadal scales, while hydrometeorologists are more interested in studying short time-scale events (i.e., hours up to a few days) such as severe storms and flash floods [6, 8].

Hydrometeorological records started in *ca* 3000 BC mainly by observing the movement of moon and stars. Since then, our understanding of hydrometeorology has advanced considerably, especially with the significant growth of technology in the second half of the 20th century (e.g., introduction of televisions in the early 1950s and computers in the 1970s).

From the 1980s up to now, tremendous advances have been made in the hydrometeorological science [9]. Governmental and private agencies have begun hiring hydrologists to use meteorological data and improve the accuracy of hydrometeorological predictions. A better knowledge of hydrometeorology along with the enhanced computational capabilities allowed them to forecast hydrometeorological variables more accurately. Fortunately, TV networks and websites have provided timely information on the weather and climate forecast.

With the advent of satellites and radars, hydrometeorology has changed from a "data poor" to a "data rich" environment [10]. Nowadays, hydrometeorologists incorporate remotely sensed data from radars and satellites into numerical models to estimate hydrologic variables such as rainfall, evapotranspiration, soil moisture, and vegetation dynamics over large-scale domains. Indeed, the technology boom and the vast amount of radar and satellite observations have enabled many national hydrometeorological centers to become hubs of information and research in the field of weather forecasting for governments, policymakers, and private agencies. The improvements over the last 50 years have been impressive, and hydrometeorological centers are continuously adopting modern technologies to provide more reliable weather and climate information for societal needs [1–4].

Five eras can be identified as the benchmark for historical advances in the science of hydrometeorology: (1) Prehistoric times (*ca* 3500–750 BC), (2) Historical to medieval times (*ca* 750 BC-1400 AD), (3) Early and mid-modern times (*ca* 1400–1800), (4) Modern times (1800–1900), and (5) Contemporary times (1900-present).

This study provides new information and insights about history of hydrometeorology. A comprehensive review of hydrometeorology in each of the abovementioned five eras contributes to a growing awareness of observational methods. As noted by the great Chinese philosopher, Confucius (*ca* 551–479 BC): *Study the past, if you would divine the future*.

2. Hydrometeorology in prehistoric times (*ca* 3500–650 BC)

In the prehistoric period, also known as the speculation period, the meteorological knowledge was based solely on speculative theories [9]. In this long era, hypotheses with no empirical validation were developed to describe meteorology, weather, and climate [9]. The prehistoric times cover the pre-Aristotelian era that is long before the invention of meteorological instruments.

The first primitive human societies began in the late Neolithic era [9]. The transition from hunting-gathering to farming increased the vulnerability of societies to climate-related hazards because they no longer migrated to avoid unfavorable environmental conditions. There was no study of meteorology at that time. Also, atmospheric phenomena could not be adequately explained. Hence, a collection of linguistic weather "signs" was created and transferred from generation to generation. For instance, moving a light or star at night was considered as a sign for sunny or rainy condition in the next day.

Figure 1.
The Nilometer at Rhoda Island in Cairo.

About 3000 BC, an ancient instrument (called a Nilometer) was first used to measure the water level of the Nile River. The Nilometer helped farmers irrigate their farms more efficiency [11]. **Figure 1** shows the Nilometer at Rhoda Island in Cairo (in 861 AD), which was designed by Afraganus.

The Babylonian king Hammurabi (*ca* 1792–1750 BC) related seasons and weather conditions to the solar cycle by developing the 360-day calendar. This calendar was used to study hydrometeorological phenomena in next centuries [11].

From 747 BC to 737 BC, other Babylonian kings (e.g., Nabu-nasir) recorded the movement and location of the moon over a period of several years. He also monitored the time of sunrises, sunsets, and eclipses. The Babylonians used this information to predict celestial events [12]. Another Babylonian king, named Nabu-suma-iskun (*ca* 700 BC), stated that a halo around the sun or the moon is a sign of flood during winter [13].

3. Hydrometeorology in the archaic to medieval times (*ca* 650 BC–1400 AD)

The theocratic explanation of meteorology was dominant until the 7th century BC. The Ionian Stoa (School) in Asia Minor was founded *ca* 600 BC by the Thales of Miletus, the father of natural philosophy and water science. The natural philosophers (the so-called pro-Socratic philosophers) such as Thales, Anaximander, Anaximenes, Pythagoras, Heraclitus, Zenos, Empedocles, Democritus, and Alcmaeans lived in Greece from the end of the 7th century until the middle of the 5th century BC. They raised new questions about the natural phenomena such as rain, cloud, storm, and lightning [14]. One of their questions was: "Is there a reality that does not change despite the ever-changing appearances of things?" [14]. While the manifestations of nature are extremely complex, the beginning (i.e., the source substance) was thought to be relatively simple. Having said that, the "beginning" was water for Thales and the air for Anaximenes [14].

In the late Archaic times, the Ionian philosophers learned fundamental hydrological processes by studying meteorological phenomena [14]. For example,

Anaximander (*ca* 610–546 BC) explained the relationship between rainfall and sunshine in his book entitled "*On Nature*". For the first time, Xenophanes (*ca* 570–475 BC) expressed the concept of the hydrological cycle and the role of sea in it.

In 465 BC, Anaxagoras (*ca* 500–428 BC) used the ideas of the Ionian philosophers to develop rain gauge instruments in Athens [15]. At that time, the first measurements of rainfall began in India [16, 17]. Later, in 100 AD, a recording rain gauge was developed in Palestine [18]. In 400 BC, Kautilya wrote a book entitled *Arthashastra*in, which elaborated the importance of rainfall for military operations [16, 17].

Plato, a well-known philosopher, advanced the concept of the hydrologic cycle by stating that "rivers and springs originate from rainfall". In 387 BC, the Platonic Academy was founded in Athens by Plato (*ca* 428–348 BC) based on the principles of the Ionian Stoa. The hydrological cycle was characterized in that Academy. Aristotle (384–322 BC) was Plato's student, and his theories were influenced by Ionian philosophers. He explained several hydrometeorological phenomena such as physics of clouds, rivers, precipitations, and changes in land covers [19–24].

In 300 BC, Theophrastus (*ca* 371–287 BC) published his Book on Signs (*De Signis Tempestatum*), which is considered to be the first weather forecasting manual. In 240 BC, Eratosthenes compared the intensity of Sun's rays at two points on the earth to calculate the spherical size of the earth and its circumference [25].

The well-known astronomer, Ptolemy (*ca* 100–170 AD), defined the earth's climatic zones on the basis of astronomical observations and air temperature variability. Recognition should also be given to two Roman scholars, Seneca (4 BC-65 AD) and Pliny (23/24–79 AD). Seneca studied a wide spectrum of meteorological phenomena (e.g., wind, lightning, thunderstorm and hurricane). Pliny collected all the meteorological theories of the Ancient Greeks [26].

Heron of Alexandria (*ca* 10–75 AD) was a physicist, mathematician, and engineer at the Museum of Alexandria. He wrote many books in his field of expertise that were used until medieval times. His most important invention was the Aeolipile, the first steam turbine [27]. He is mostly known for his profound comprehension of physics, which is reflected in the Pneumatica (his description of how mechanical devices operate by air, water, and steam) [28].

Documents from the Jewish tradition (called the Mishnah) show the rain water harvesting practice from *ca* 200 BC to 200 AD. During the Han dynasty (*ca* 206 BC-220 AD) in China, the hydrological cycle was represented by the 1) water vapor transfer from the land surface to the overlying atmosphere due to evaporation, and 2) cloud formation.

In 800 AD, Vikings in Scandinavia believed in Thor as the god of thunders and lightning, and Freyr as the god of sun, rain, and other meteorological phenomena. Thunder was the sound of Thor, and lightning was the sign of killing Thor's enemies [29].

In *ca* 1000 AD, Ibn Wahshiyya discussed the importance of weather forecasting for agricultural production in his book entitled "*Nabataean Agriculture*". According to Wahshiyya, a visible moon is a sign of clear weather in the next day. Unlike the ancient Babylonians, he believed that clear weather would come if the moon was surrounded by a halo. Wahshiyya also mentioned that thin (thick) clouds were the sign of cold (warm) weather. Based on his beliefs, an owl's hoot implies the closeness of cold weather [30].

In 1328 AD, William of Ockham (1285–1347 AD) wrote a great deal on natural philosophy and attempted to quantify atmospheric physics and other natural sciences. William highlighted the importance of reliable meteorological observations in a long commentary on Aristotle's Physics [31–33].

4. Hydrometeorology in early and mid-modern times (*ca* 1400–1800)

During this era, many scientists tried to develop new methods and instruments to monitor hydrometeorological variables. The end of the era of theories and the beginning of the modern meteorology are demarcated by the first 'modern philosopher', Rene Descartes (1596–1650 AD). He established the principle of scientific philosophy in his work *Les Meteors,* denoting nothing should be accepted as truth unless it is proven [34].

The economy of Korea in the Far East during the Joseon Dynasty (1392–1897) was mainly dependent on agriculture. Thus, Koreans had to manage their water resources efficiently. Jang Yeong-sil designed the first Korean rain gauge (called *cheugugi*) in 1441 (**Figure 2**). In 1442, the standard rain gauges with the height of 42.5 cm and diameter of 17 cm were installed across Korea to record rainfall data [9].

As the Renaissance began, weather forecasts were based on astrology and interpretation of weather signs. Meteorological instruments were improved only slightly from the middle ages until the beginning of age of instrumentation in the 17th century. In 1450, Nicholas of Cusa developed an idea for a hygroscopic hygrometer to measure air moisture. His plan was to use wool and stones on different sides of a large scale. The hygrometer operated based on the ability of wool fibers to absorb air moisture. In 1481, Leonardo DaVinci took advantage of Nicholas' idea and made the first hygrometer (**Figure 3a**). DaVinci's invention was used until 1500 [9]. Later, Francesco Folli (1624–1685) created a hygrometer (**Figure 3b**), which he named as *"Mostra Umidaria"* (in Italian). In his hygrometer, a frame carries a small roll at each end, on which is wrapped the end of a paper ribbon (now missing) serving as a hygroscopic substance. The frame is made of brass and has the shape of a finely decorated balustrade. The center of the frame holds a decorated brass dial fitted with a circular graduated scale. By means of a simple mechanical system, the dial indicates the changes in ribbon length due to the variations in atmospheric humidity [37]. Nevertheless, the development of hygrometer as a scientific instrument was started in 1768 by the German mathematician John Heinrich Lambert (*ca* 1728–1777).

The British physicist, Robert Boyle (1627–1691), was one of the first scientists that recognized the need for a standard thermometric scale to make temperature measurements comparable. In 1714, Gabriel Daniel Fahrenheit (1686–1736) built a mercury thermometer that could measure the temperature as low (high) as the freezing (boiling) point of water. Anders Celsius (1701–1744) proposed a new scale for thermometers in 1742. His 'centesimal' (meaning 100 divisions) system was easier to use in scientific works and became the basis for the 'Celsius' or 'centigrade' temperature scale. In 1862, Lord Kelvin (1824–1907), the Scottish mathematician and physicist, used the absolute zero (or zero degree Kelvin) in the so-called absolute temperature scale. Absolute zero is defined as the temperature in which molecules stop moving [38]. Benedetto Castelli (1578–1643) built the first rain gauge in the 16th century in Italy. More meteorological instruments were developed by other Italian meteorologists. Ferdinand II de' Medici (1610–1670) built a condensation hygrometer that operated by exposing water vapor to a cylindrical iced glass.

The science of hydrometeorology has benefited from advances in mathematics and physics. The relationship between air pressure and height was one of the most interesting subjects in the history of hydrometeorology. Evangelista Torricelli and Blaise Pascal developed the first barometers in 1643 and 1646, respectively. In 1660, Robert Boyle found the relationship between the gas pressure and volume. Benjamin Franklin (1706–1790), the American statesman and scientist, discovered the electrical nature of lightning in 1752. He also realized that storms can move from

Figure 2.
The first Korean rain gauge designed by Jang Yeong-sil [9].

a b

Figure 3.
(a) The DaVinci's hygrometer [35, 36], and (b) the Francesco Folli's hygrometer, named as Mostra Umidaria [37].

place to place. In 18th century, the first studies on dynamic meteorology were done by Halley and D'Alambert [39].

The equations of motion, the continuity equation, the first law of thermodynamics, the state equation of gases (the law of ideal gases), and the hydrostatic equation have been used to describe atmospheric motions [9]. The first law of thermodynamics was formulated in the 19th century by Germain Hess and Rudolf Clausius. Isaac

Newton (1642–1727) introduced many principles of mechanics in an integrated framework and highlighted their use in describing atmospheric phenomena [40]. Leonard Euler (1707–1783) investigated the variation of air pressure with height above sea level. Using the Newton's second law of motion, he developed the equations of fluid flow in 1755, which were a significant contribution to fluid mechanics [40].

The development of hydrometeorology was further enhanced in the late 17th and early 18th century. Robert Hooke (1635–1703 AD) had new ideas for designing hydrometeorological instruments. His collaboration with Sir Christopher Wren led to the construction of the first automatic rain gauge, called the tipping bucket rain gauge [27, 33]. Richard Towneley (1677–1704 AD) used their automatic rain gauge to measure rainfall in the UK [27, 33].

In the 18th century, scientists improved the accuracy of instruments to monitor meteorological phenomena more reliably. In 1743, Benjamin Franklin studied the movement of hurricanes in Philadelphia and Boston. Horace-Benedict de Saussure (1740–1799 AD) improved the accuracy of hygrometers by designing a hair hygrometer in 1775, which is still used today [41, 42]. de Saussure's hair hygrometer works based on changes in the length of a human hair as air humidity varies. In the late 1700s, hydrometeorologists started to monitor meteorological variables over large-scale areas. For instance, in 1777, David Dobson measured raindrop size and evaporation over Liverpool in the UK ([29, 43]). In summary, the 18th century was characterized by the development of basic meteorological instruments and dynamic equations [44]. However, the widespread application of these tools and equations began in the 19th century.

5. Hydrometeorology in modern times (1800–1900)

In the 19th century, meteorological scientists used new devices (e.g., psychrometers, hygrometers, meteorographs, and weather kites) in weather stations. They also categorized several meteorological phenomena such as clouds, hurricanes, and tornadoes. The English naturalist Luke Howard (1772–1864) classified different types of clouds in 1803. In 1830, the Connecticut merchant William Redfield (1789–1857) discovered the circular motion of hurricanes. He also classified different types of hurricanes and tornadoes [44, 45].

The British Admiral Francis Beaufort (1774–1857) developed a wind force scale for mariners in 1806. Ernst Ferdinand August (1795–1870) built a psychrometer to measure air humidity in 1818. It used a dry bulb thermometer and a wet bulb thermometer to measure air temperature [9]. The difference in temperature of the two thermometers was utilized in the Ernst Ferdinand August's algorithm to obtain air humidity. In 1820, John Frederic Daniell (1790–1845) invented a new type of hygrometer, called a dew point hygrometer [46]. He cooled down a polished metal mirror to a temperature at which water vapor in the air began to condense on it (i.e., dew point temperature). William Jevons (1835–1882) corrected the errors of rain gauge measurements due the wind. George James Symons (1838–1900) expanded Jevons' corrections and founded the British Rainfall Organization (British [47, 48]).

The first synoptic weather charts were constructed by the German meteorologist Heinrich Wilhelm Brandes (1777–1834) in 1820. These weather charts were a significant milestone in the history of theoretical and applied meteorology. These weather charts indicated the low- and high-pressure systems and initialized the field of synoptic meteorology [49, 50].

By the invention of telegraph in 1843, the first weather observation network was established in the UK to transmit weather observations to various stakeholders. In 1860, the Met Office in London started to transmit meteorological measurements to the community by telegraph. The British Meteorological Society (BMS) was created in 1850 and later renamed to the Royal Meteorological Society (RMS). In 1861, the Met Office began publishing weather forecasts for the public in the UK [51, 52].

The first attempt to automatically record meteorological variables was made by Father Secchi in 1867. He designed the first Meteorograph to measure air pressure and rainfall duration (**Figure 4a**). In 1870, weather balloons were employed by Alexander Wilson in the UK to collect meteorological data [9]. In the late 1880s, meteorological stations were installed in other countries. For example, the civic association Stadtverein Salzburg installed a weather station in Salzburg (Austria) in 1888 to measure air pressure, temperature, and humidity (**Figure 4b**) (Atlas [54]).

In 1870, President Ulysses S. Grant established a weather bureau in the US, currently called the National Weather Service (NWS) [55]. By the 1870s, the US had more than 20 weather stations that transmitted micrometeorological data to Washington DC by telegraph.

The International Meteorological Organization (IMO) was founded in 1873 to facilitate the cooperation among all national weather services. The IMO organized several international meteorological conferences in Vienna, Rome, Munich, and Paris in 1873, 1879, 1891, and 1896, respectively. The IMO was renamed to the World Meteorological Organization (WMO) in 1950. Nowadays, the WMO serves as the specialized agency of the United Nations for meteorology (weather and climate), agrometeorology, operational hydrology, and related geophysical sciences [9]. Currently, the

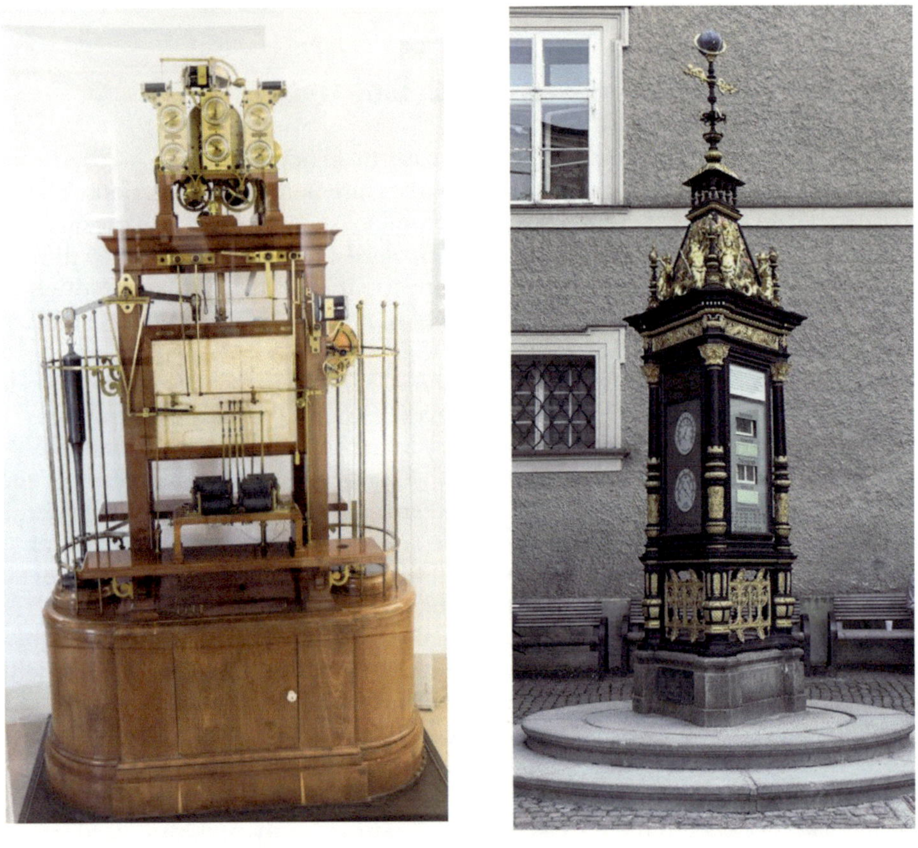

a b

Figure 4.
(a) The Meteorograph designed by father Secchi [53], and (b) the first (19th century) weather station in Salzburg, Austria.

DOI: http://dx.doi.org/10.5772/intechopen.94939

Figure 5.
Measuring weather information by a weather kite in 1894. The location of the picture is unknown [9].

WMO has at least 187 member states and territories. The Japan Meteorological Agency (JMA) and the Meteorological Society of Japan were formed in the 1880s [9].

In the late 1890s, further attempts were made to measure various micrometeorological variables. Weather kites were used in 1894 to collect air temperature, pressure, humidity, and wind speed at high altitudes (**Figure 5**) [9]. Weather kites were used instead of weather balloons (developed in late 1780s) as they could move more readily. In 1898, the Richard brothers in France invented a barothermograph, which consisted of a thermometer, a barometer, and a hygrometer [56].

6. Hydrometeorology in contemporary times (1900–present)

The modern hydrometeorology was born in contemporary times (1900-present). In this era, weather data were used in the forecast models [9].

The discovery of the stratosphere at the beginning of the 20th century by Léon Philippe Teisserenc de Bort (1855–1913) advanced meteorology and hydrometeorology. Similarly, the discovery of the tropopause (i.e., the buffer zone between troposphere and stratosphere) by Ernest Gold (1881–1976) and William Jackson Humphreys (1862–1949) in 1900 further augmented hydrometeorology. In 1920, the Norwegian School of Meteorology (NSM) made a significant contribution to the field of meteorology by organizing seminars and inviting the most well-known scientists to them [57]. New theories about frontal surfaces and development of low-pressure systems by the Norwegian scientists, namely Vilhelm Bjerknes (1862–1951), Jacob Aall Bonnevie Bjerknes (1897–1975), Halvor Solberg (1895–1974), and Tor Bergeron (1891–1977) took a dominant position in hydrometeorology [58]. In 1940, Carl Gustaf Rossby (1898–1957) from the US weather service discovered the jet stream and its controls over the easterly movement of most weather systems [59].

Hugo Hildebrand Hildebrandsson (1838–1925) published his book entitled *International Cloud Atlas.* His book improved meteorologists' knowledge of cloud physics [60, 61]. In 1909, the Met Office equipped ships with wireless telegraphy to transmit weather records to designated centers in the UK. This was the first attempt to transfer real-time weather data from the ships to the land [62]. The American Meteorological Society, AMS (founded in 1919), has advanced our understanding of meteorology, hydrometeorology, and hydroclimatology. In 1922, Lewis Fry Richardson used weather observations in his simple numerical model to forecast air pressure and wind speed. Today, complex numerical models are used instead of his simple weather forecast algorithms [9, 63].

In 1922, the first weather radio broadcasts were developed in New York and London [64]. In 1921, the Hydrometeorological Center of Russia was founded in Moscow. Nowadays, this center has more than 30 laboratories, departments and administrative branches, and provides forecasts of hydrometeorological variables. The US broadcasted the first weather forecast program on television in 1941 [64].

The prestigious seminars of Norwegian School of Meteorology, the invention of radiosondes, the Bergeron's theory of rain formation, the ionospheric research of Edward Victor Appleton (1892–1965) and Miles Aylmer Fulton Barnett (1901–1979), and the theoretical studies of low- and high-pressure systems were conducted between the First and Second World Wars (1920–1940). In addition, in this period, there were some studies on the general circulation of atmosphere, the properties of motion, the mechanisms of fronts and low-pressure systems, the atmospheric disturbances, and the isentropic analysis [58]. During the Second World War, radiosonde meteorological measurements led to the discovery of jet stream. In 1935–1945, new instruments such as weather radars and radio wave sensors were invented [58]. In the same period, several studies were performed on the chemical composition of the upper atmosphere and fog decomposition [58].

Jule Charney used the first computer in 1950, called the Electronic Numerical Integrator and Computer (ENIAC), to run his meteorological model [9]. In the 1950s, the first computational atmospheric models were developed and used in hydrometeorology. In this decade, government agencies took advantage of geographical information to forecast weather more accurately. In 1954, the first radar weather station was built in New Orleans, USA [9]. In 1959, the Met Office created a computer, called Meteor, which was able to conduct 30,000 calculations per second [65].

The chaotic nature of the atmosphere was first realized by Edward Norton Lorenz (1917–2008) in the 1960s. He introduced the chaos theory and limitations of atmospheric predictability. This is also known as the butterfly effect as flapping of a butterfly's wings can cause a large disturbance somewhere else [58]. The first generation of satellites emerged in this decade. In 1960, the first weather satellite, called the Thermal Infrared Observation Satellite (TIROS), was launched by the US. This satellite could send 4000 images per week to the earth [66].

The National Aeronautics and Space Administration (NASA) lunched the Synchronous Meteorological Satellite-1 (SMS-1) and SMS-2 in the 1970s. Other geosynchronous meteorological satellites were also launched by the NASA as part of the Geostationary Operational Environmental Satellite (GOES) program.

There is a large number of satellites at present, which is steadily increasing year by year for remote sensing of the earth. Remote sensing can be classified into two main categories: (1) active radars, and (2) passive instruments (sensors). The active radars transmit energy and record the backscattered signals. Weather radars can operate even in cloudy skies because their signals can pass through clouds. The passive systems record the emitted radiation from the earth.

Scientific and technological advances led to the development of Doppler and dual-polarization weather radars, which are currently used to detect storms [44]. These radars allow researchers to "see" inside the storms and monitor wind-driven precipitation. They also visualize the wind rotation and allow meteorologists to detect severe storms such as tornadoes and mesoscale convective systems.

Meteorological satellites are located in either low polar (e.g., Polar Orbiting Environmental Satellites (POES) and Television Infrared Observation Satellites (TIROS-N)) or high geostationary (e.g., meteorological satellites (METEOSAT) and GOES satellites) orbits. A number of widely used meteorological satellites are shown in **Figure 6**. They monitor the weather, soil moisture, sea and land surface temperatures, precipitation, crop condition, snow depth, land cover, landslide, etc. [16]. While these satellites allow to monitor various hydrologic variables, they have main challenges regarding community acceptability, underestimating total precipitation due to light rainfall events, unquantified uncertainty, data continuity, sensor changes, and data maintenance [67, 68].

Continuity of data is crucial in order to develop reliable and accurate satellite records for hydrologic applications. Most satellites are functional for less than 10 years, though many of them operate beyond a decade. Although, launching satellites should be extended for follow-up missions, designing satellites require substantial investments and can take decades. The Global Precipitation Measurement (GPM) and Gravity Recovery and Climate Experiment (GRACE) are two examples of satellite missions planned to fix the problem of the gaps in the current satellite-based precipitation and total water storage data, respectively [67, 68]. Nowadays, the quality and resolution of satellite images are significantly improved [69, 70]. **Figure 7** shows the image of the Gulf of St. Lawrence from the TIROS-1 weather satellite (1970s) and Suomi National Polar-orbiting Partnership (S-NPP) satellite. As can be seen, there is a remarkable improvement in the quality and resolution of the image from 1970s up to 2013. The improvement in satellites has helped understand hydrological phenomena such as glacial lake outburst flood [71] and soil erosion [72], which were difficult to study in the past [71].

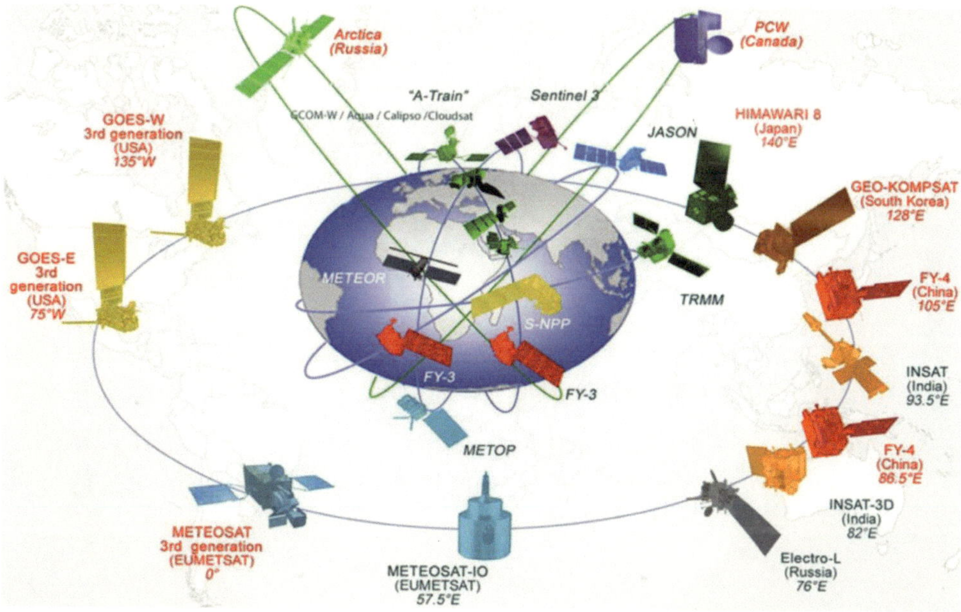

Figure 6.
A number of widely used satellites that monitor and transit meteorological and climatological information [9].

Today, new technologies such as microwave sensors (**Figure 8a**) and drones (**Figure 8b**) allow to monitor extreme events (e.g., floods, droughts, and hurricanes) and mitigate their damage on the environment, infrastructures, and critical resources [9].

Weather models were developed at the end of 20th century and beginning of 21st century. The first real-time medium-range forecasting model was developed by the European Centre for Medium-Range Weather Forecasts (ECMWF) in 1979. The Intergovernmental Panel on Climate Change (IPCC) was founded in 1988 by the United Nation (UN) to monitor climate change, and its economic, social and environmental impacts across the world. In the 1990s, the Weather Research and Forecasting (WRF) model was developed by simulating the atmospheric processes. This model has been used in more than 150 countries around the world to simulate the atmosphere via real-time data [9]. In 2002, the Aviation Model (AVN) was developed by the National Centers for Environmental Prediction (NCEP) for short-range weather forecasting. This model (called the Global Forecasting System, GFS) is the leading forecasting model in the US [69]. A review of history of hydrometeorology is provided in Appendix A as supplementary materials.

Figure 7.
The image of the Gulf of St. Lawrence from the (left) TIROS-1 weather satellite (1970s) and (right) S-NPP satellites (2013) [9].

a b

Figure 8.
(a) A monitoring site which is equipped with microwave sensors to provide real-time measurements of water level for forecasting storm surge [9], and (b) NASA's drone (Sierra) for remote sensing sampling in inaccessible regions such as polar regions, mountaintops, and open waters [70].

7. Emerging trends

Recent advances in active and passive remote sensing systems have created new cost-effective opportunities for meteorological applications. The new generation of satellites (e.g., Soil Moisture Active Passive (SMAP), Meteosat Third Generation (MTG), Himawari-9, and FY-4B satellites) allows monitoring the earth and atmosphere with higher spatial and temporal resolution. Progresses in technology have also improved field instruments used to collect weather data. In addition, numerical models have become more advanced as new theories and concepts were incorporated into them, allowing them to simulate hydrometeorological processes more accurately. At present, three-dimensional coupled atmosphere–ocean models use remotely sensed and in-situ observations to forecast weather up to ten-days ahead [9].

Weather modification by cloud seeding and aerosols spreading is one of the emerging trends to decelerate the threats associated with global warming [73]. The main purpose of cloud seeding and aerosols spreading is to alter rainfall patterns [73]. However, these methods are expensive and have not yet reached a practically acceptable level. In addition, the amount of precipitation that reaches the land surface is often negligible. This happens because snowfalls and/or light drizzles generated by the stratiform clouds are evaporated prior to reaching the ground level. To overcome this issue, stratocumulus clouds should be created that can have a global impact [74]. On the other hand, a number of scientists hypothesize that the weather modification can cause climate change and may lead to extreme weather events such as drought and flood [73, 75]. Some hydrometeorologists believe that the chemicals used in cloud seeding are dangerous for human health because of their toxicity and damage the ozone layer [75]. Moreover, the increase of particulate matters in the weather modification process may change the color of the sky from blue to gray [75].

Applying the science of hydrometeorology to real-world problems is another emerging trend. It is called operational hydrometeorology and deals with the application of hydrometeorology to real-time operational systems. The major components in an operational prediction system are monitoring equipment, meteorological and hydrological forecasting models, demand (water supply) prediction models, and decision support tools [76, 77]. The difficulty of forecasting rainfall has become a main challenge in the development of operational hydrometeorology [76]. There have been some attempts to overcome this issue. For example, the Flood Forecasting Center (FFC) was established in England and Wales in 2009 to forecast rainfall. A number of scientists used inverse modeling to predict rainfall [78–81]. For example, they assimilated river discharge observations within an ensemble data assimilation framework to predict rainfall [82–84]. Similarly, several studies assimilated soil moisture observations into water balance models to improve rainfall predictions [85–91].

Using low-cost sensors for flash flood forecasting [78], and application of hydrometeorology in marine sciences [92] and urban environment [80, 81] are other emerging trends.

8. Future issues and challenges

About 90% of disasters in the world during 1995–2015 were related to weather [93]. During this period, more than 600,000 people died, and more than 4 billion others were evacuated or injured because of weather-related hazardous events. The annual cost of damages caused by weather and climate extremes at the global scale is about $300 billion. Most disasters have been observed in the US, China, India, Philippines, and Indonesia [93].

Modern technologies such as advanced weather balloons, radars, satellites, and mathematical and numerical models have allowed to mitigate the impact of weather-related disasters on human beings and environment. In addition, innovative hydrometeorological devices and synoptic stations have provided concrete weather data to further lessen the effect of extreme weather events.

Despite these advances, the complexity of climate requires the development of more accurate models and instruments to manage the natural disasters more efficiently [9]. Overall, the future of the science of meteorology and hydrometeorology relies on new sophisticated instruments and prediction models, which enhance our ability to forecast weather and mitigate related hazards [9]. Having said that, the fourth industrial revolution (IR 4.0) may help the science of hydrometeorology by developing microchips, microcontrollers and more accurate sensors (i.e., multi-sensor meteorology) that can be utilized in weather sites [94–99].

Today, hydrometeorologists take advantage of satellite data at different spatial and temporal scales [100, 101]. Artificial intelligence (AI) and machine learning (ML) approaches can use long-term remotely sensed data from satellites to improve weather prediction and climate modeling capabilities [102–104]. Also, the advancement in Internet of Things (IoT) will make real-time data observations more precise [105–108].

9. Conclusions

This study provides a thorough review of the historical evolution of the science of hydrometeorology and its significant milestones from past civilizations to contemporary times. Hence, it can expand our knowledge of the advances in hydrometeorology through different centuries. In the past civilizations, the first steps were taken to understand weather changes. Today, the availability of robust numerical models, remote sensing data, and high computational capabilities have allowed humankind to predict meteorological and climatological events. Five major periods are considered in this study: 1) the prehistoric, 2) the archaic and medieval, 3) the early and mid-modern, 4) the modern, and finally 5) the contemporary periods. The key advancements and achievements in each period are presented.

The theocratic explanation of meteorology was dominant until the 7th century BC. In the prehistoric period, weather was unpredictable. Also, religion, folklore, tradition, culture, and beliefs were the main elements for studying hydrometeorology. In the late Archaic times, the Ionian philosophers explained hydrometeorological processes for the first time. Beginning in the historical period, Anaxagoras (*ca* 500–428 BC) used the ideas of the Ionian philosophers to develop rain gauge instruments in Athens. Also, in this period, the first evidence of measuring rainwater was seen in Greece and India. Later, Plato (*ca* 428–348 BC) developed the concept of the hydrological cycle in his academy in Athens. In the early Hellenistic times, Theophrastus of Eresos (*ca* 371–287 BC) wrote the book *Signs De Signis Tempestatum*, which was the first weather forecasting manual.

From 27 BC to 200 AD, Pomponius Mela, the Roman Emperor in Spain, worked on geographical maps and divided the earth into five climate zones. Investigating weather and atmospheric phenomena was almost stopped from the end of the Roman period to the Middle Ages of the Renaissance. However, there were considerable attempts from *ca* 1400 to 1900 AD to monitor hydrometeors and forecast weather by meteorological instruments, which were invented during this period.

From 1950 until present, theoretical approaches and mathematical analyses have been extensively used in the science of hydrometeorology. Sophisticated instruments have been developed to measure hydrometeorological variables. Computers have been utilized to solve complex mathematical equations and run numerical models to understand meteorological phenomena in the light of the application of meteorological theories (e. g., the application of heat and mass transfer theories to analyze evaporation).

The development of research and education in the field of hydrometeorology began after the Second World War, and accelerated with the formation of the World Meteorological Organization (WMO) in 1951. Scientists in the modern era have provided foundations for hydrometeorological investigations and instrumentations in a universal scale. Their efforts have improved humans' understanding of atmospheric phenomena.

The perspectives in the field of hydrometeorology are promising. This is mainly due to the advances in sensors and instrumentation, computational capabilities, remote sensing systems, data mining techniques, information and communication technologies (ICTs), decision support systems (DSS), and deep learning approaches. Although the science of hydrometeorology has significantly improved recently, there is still lack of adequate knowledge to accurately forecast extreme hydrometeorological events.

Appendix

Prehistoric times (*ca* 3500–750 BC)	
Ca 3500 BC	"Astrometeorology" emerged in Babylon. The sensitivity of humans to weather increased because they no longer migrated
Ca 3500 BC	Early Egyptians established sky-religion and rainmaking rituals
Ca 3000 BC	Nilometers were used to record water levels in the Nile River
Ca 1800 BC	Nilometers were developed at the second cataract of the Nile River
Ca 1750 BC	Water codes of King Hammurabi (*ca* 1792–1750 BC), which consisted of 282 regulations
Ca 740 BC	Nabu-nasir (*ca* 747–734 BC) regularly recorded movement and location of the moon. He also noted the times of sunrises, sunsets, and eclipses
Historical to medieval times (*ca* 750 BC-1400 AD)	
Ca 600 BC	The Thales of Miletus (*ca* 624–546), the founder of Ionian Stoa (School), is considered to be the father of natural philosophy and water science. He introduced the hydrologic cycle. He also presented a physical exegesis for the Nile flooding during summer time when rainfall in Egypt was minimal.
Ca 570 BC	Anaximander (*ca* 610–546 BC) explained the relationship between rainfall and evaporation in his book entitled "*On Nature*". The first known work on the natural philosophy.
Ca 550 BC	Anaximenes (585–528 BC) explained the formation of winds, clouds, rainfalls, and hails
End of *ca* 5th BC	Xenophanes (*ca* 570–475 BC) expressed the concept of hydrological cycle and the role of sea in it.
Ca 500 BC	First attempts to measure rainfall in Greece
Ca 465 BC	Anaxagoras (*ca* 500–428 BC) transferred the ideas of the Ionian philosophers to the Athenians. He also explained the formation of hailstorms.
Ca 400 BC	The first measurements of rain fall in India
Ca 400 BC	Hippocrates of Cos (*ca* 460–370 BC) studied the effects of climate and environment on human health in his treatise on *Airs, Waters, and Places*
Ca 387 BC	The Platonic Academy was founded by Plato (*ca* 428–348 BC). The concept of hydrological cycle was developed in that academy

Ca 345 BC	Aristotle (*ca* 384–322 BC) founded the Lykeion of Aristotle, also known as the "*Peripatec School*"
Ca 340 BC	Aristotle summarized his meteorological knowledge in his book entitled *Meteorologica*.
Ca 332 BC	Alexandria was founded in a small ancient Egyptian town by Alexander the Great
Ca 330 BC	Theophrastus of Eresos, Lesbos (*ca* 371–287 BC) Book on *Signs De Signis Tempestatum* is considered as the first weather forecasting manual
Ca 300 BC	Theophrastus On Winds (De Ventis) accepted the Presocratic's hypothesis of wind's origin. He also introduced a basic understanding of atmospheric pressure
Ca 250 BC	Archimedes (*ca* 287–212 BC) explained the buoyancy principle
Ca 240 BC	Eratosthenes (*ca* 276–194 BC) reported that the earth is a globe with the circumference of 40,000 km
Ca 240 BC	Philo of Byzantium (*ca* 280–220 BC) invented a device that measured the expansion and contraction of air as it warmed up and cooled down, respectively.
Ca 25 AD	In Spain, the Pomponius Mela, Roman Emperor introduced the climate zone systems
Ca 60 AD	Hero (Heron) of Alexandria (*ca* 10–75 AD) is mostly known as an engineer and designed a basic thermometer. Also, his treatise *Pneumatica* (Pneumatics) advanced the science of physics
Ca 70 AD	In Rome, Gaius Pliny Secundus (Pliny the Elder) (ca 23/24–79 AD) developed the encyclopedic Natural History, which later became an editorial version for encyclopedias
Ca 200 AD	In Tunisia, the Quintus Septimus Florens Tertullianus (160–225 AD) ended the observation-based science, and began the "sacred science" based on the "authority" of scripture
Ca 380 AD	Based on the prophecies of Isaiah and the Epistle to the Ephesians, St. Jerome (*ca* 347–420 AD) considered a doctrine of the diabolical origin of storms.
Ca 400 AD	In Algeria, St. Augustine, Bishop of Hippo (354–430 AD) whole heartedly supported the diabolical origin of storms.
Ca 900 AD	Chinese weighted charcoals to measure the air moisture
Ca 1000 AD	Ibn Wahshiyya translated the book entitled "*Nabataean Agriculture*". The importance of weather forecasting for agriculture was discussed in this book.
Ca 1247 AD	Gauges (made of large bamboo segments) were used to measure precipitation in China
Ca 1328 AD	William of Ockham (1285–1347 AD) attempted to advance natural sciences and atmospheric physics by improving the quality of observations.
Ca 1300 AD - 1400 AD	Air temperature (1400 BC) and rainfall (1216 BC) were recorded in ancient China
Early and mid-modern times (*ca* 1400–1800)	
Ca 1442	A simple cylindrical container was used to collect precipitation in Korea
Ca 1450	Leon Battista Alberti invented a flat plate anemometer in Italy
Ca 1450	Nicholas of Cusa invented a hygroscopic hygrometer
Ca 1500	Leonardo Da Vinci (1452–1519) improved the hygrometer, which was developed by Nicholas of Cusa
Ca 1593	Galileo Galilei (1564–1642) invented a thermometer in Italy
Ca 1639	Benedetto Castelli (1578–1643) constructed the first scientific rain gauge in Italy and Europe

Ca 1643	Evangelista Torricelli (1608–1647) invented the barometer
Ca 1648	Blaise Pascal (1623–1662) invented a barometer based on variations of atmospheric pressure with altitude
Ca 1660	Francesco Folli (1624–1685) created a paper-ribbon hygrometer, called *Mostra Umidaria*
Ca 1663	Robert Hooke (1635–1703) collaborated with Sir Christopher Wren to build the first automatic rain gauge called tipping bucket rain gauge. However, the first measurements of rainfall were done by Richard Towneley (1677–1704 AD)
Ca 1665	Grand Duke Ferdinand II de' Medici (1610–1670) created the condensation hygrometer
Ca 1667	Robert Hooke invented the anemometer
Ca 1670	Robert Hooke invented the first mercury glass-thermometer
Ca 1675	Horace-Benedict de Saussure (1740–1799) created the first hair hygrometer
Ca 1687	Isaac Newton (1643–1727) detailed his three laws of motion
Ca 1743	Benjamin Franklin (1706–1790) realized the northeastward movement of a hurricane from eclipse observations at Philadelphia and Boston.
Ca 1777	David Dobson developed the ideas to measure evaporation and raindrop size
Modern times (1800–1900)	
1818	Ernst Ferdinand August (1795–1870) developed ideas to create psychrometer
1820	John Frederic Daniell (1790–1845) invented a new hygrometer, called dew point hygrometer
1850	The British Meteorological Society was established and then renamed to Royal Meteorological Society (RMS)
1860	Meteorological observations were being made routinely by the Met Office in London
1861	William Jevons (1835–1882) reduced errors in rainfall measurements using a wind shield for rain gauges
1861	The Met Office began reporting weather forecasts for the public in England
1867	Father Secchi invented the first Meteorograph
1870's	Weather observations from 20 stations were transmitted to Washington DC via telegraph
1870	President Ulysses S. Grant established a weather bureau, which is now called the National Weather Service (NWS)
1870	Alexander Wilson used weather balloons in the UK to collect weather information
1873	The International Meteorological Organization was formed. It is now named the World Meteorological Organization (WMO) and is an entity of the United Nations
1879	George James Symons (1838–1900) expanded Jevons' theory and founded the British Rainfall Organization
1880's	The Meteorological Society of Japan were formed
1894	Weather kits were used to collect air temperature, pressure, humidity and wind speed at higher altitudes
1896	Léon Teisserenc de Bort (1855–1913), a French meteorologist, used weather balloons to measure air temperature and humidity
1898	The Richard brothers of France invented barothermograph

Contemporary times (1900-present)	
1907	Hugo Hildebrand Hildebrandsson (1838–1925) published his book entitled *International Cloud Atlas* and developed seasonal forecasts of clouds
1920's	Concepts of air masses and fronts were formulated by the Norwegian meteorologists. They developed a theory for the evolution of mid-latitude cyclones, which is still in use today
1921	Hydrometeorological Center of Russia was formed in Moscow
1921	The first weather radio broadcasts were made in the US
1922	Lewis Fry Richardson used numerical methods to forecast air temperature and humidity
1941	The US broadcasted the first TV program on weather forecast
1950	Jule Charney ran his meteorological algorithms by a computer called the Electronic Numerical Integrator And Computer (ENIAC)
1954	The first radar weather station was built in New Orleans, the US
1959	The Met Office created a computer (called Meteor), which was able to conduct 30,000 calculations every second
1960	The first weather satellite, called the Thermal Infrared Observation Satellite (TIROS), was launched by the US
1970's	NASA lunched geosynchronous weather satellites
1979	The first real-time medium-range forecasting model was developed by the European Centre for Medium-Range Weather Forecasts (ECMWF)
1988	The Intergovernmental Panel on Climate Change (IPCC) was founded by the United Nation
1990's	The NWS was modernized. The Weather Research and Forecasting (WRF) Model was developed.
2002	The Aviation Model (AVN) was created for short-range weather forecasting. This model, called the Global Forecasting System (GFS), is the leading forecasting model in the US.
2015	A new generation of supercomputers with the ability to perform over 10,000 trillion calculations per second was developed.

Table A.1.
Milestones in hydrometeorology in the (1) prehistoric times (ca 3500–750 BC), (2) historical to medieval times (ca 750 BC-1400 AD), (3) early and mid-modern times (ca 1400–1800), (4) modern times (1800–1900), and (5) contemporary times (1900-present).

Author details

Mohammad Valipour[1]*, Sayed M. Bateni[1], Nicolas R. Dalezios[2], Mansour Almazroui[3], Essam Heggy[4,5], Zekai Şen[3] and Andreas N. Angelakis[6,7]

1 Department of Civil and Environmental Engineering and Water Resources Research Center, University of Hawaii at Manoa, Honolulu, HI 96822, USA

2 Department of Civil Engineering, University of Thessaly, Volos, Greece

3 Center of Excellence for Climate Change Research/Department of Meteorology, King Abdulaziz University, Jeddah 21589, Saudi Arabia

4 Viterbi School of Engineering, University of Southern California, Los Angeles, CA 90089, USA

5 Jet Propulsion Laboratory, California Institute of Technology, Pasadena, CA 91109, USA

6 HAO-Demeter, Agricultural Research Institution of Crete, 71300, Iraklion, Greece

7 Union of Hellenic Water Supply and Sewerage Operators, 41222, Larissa, Greece

*Address all correspondence to: valipour@hawaii.edu

References

[1] Bruce JP, Clark RH. Introduction to Hydrometeorology: Pergamon International Library of Science, Technology. Amsterdam, Netherlands: Engineering and Social Studies. Elsevier; 2013

[2] Houser P. Hydroclimatology and hydrometeorology. International Encyclopedia of Geography: People, the Earth, Environment and Technology: People, the Earth, Environment and Technology. 2016;**1-13**

[3] Peck EL. Hydrometeorology. Bull. Am. Meteorol. Soci. 1978;**59**(5):609-612. DOI: 10.1175/1520-0477-59.5.609

[4] Rakhecha P, Singh VP. Applied Hydrometeorology. Berlin, Germany: Springer Science and Business Media; 2009

[5] Betts AK. Understanding hydro-meteorology using global models. Bull. Am. Meteorol. Soci. 2004;**85**(11):1673-1688. DOI: 10.1175/BAMS-85-11-1673

[6] Collier CG. Hydrometeorology. Hoboken, NJ: John Wiley and Sons; 2016

[7] Fry JL, Graf HF, Grotjah R. The Encyclopedia of Weather and Climate Change. Los Angeles: University of California Press; 2010

[8] Shuttleworth WJ. Terrestrial Hydrometeorology. Hoboken, NJ: John Wiley and Sons; 2012

[9] Teague KA, Gallicchio N. The Evolution of Meteorology: A Look into the Past, Present, and Future of Weather Forecasting. Hoboken, NJ, USA: John Wiley and Sons; 2017

[10] Margulis SA, Wood EF, Troch PA. The terrestrial water cycle: Modeling and data assimilation across catchment scales. Journal of Hydrometeorology. 2006;7(3):309-311. DOI: 10.1175/JHM999.1

[11] Krasilnikoff J, Angelakis AN. Water management and its judicial contexts in ancient Greece: A review from the earliest times to the Roman period. Water Polic. 2019. DOI: 10.2166/wp.2019.176

[12] Teresi D. Lost Discoveries: The Ancient Roots of Modern Science: From the Babylonians to the Maya. New York: Simon and Schuster; 2002

[13] Verderame L. The halo of the moon. In: Fincke JC, editor. Divination in the Ancient near East. Winona Lake, IN: Eisenbraun; 2014

[14] Burnet J. Early Greek Philosophy. 3rd ed. London, UK: A and C Black; 1920

[15] Angelakis, A. N., Asano, T., Bahri, A., Jimenez, B. E., and G. Tchobanoglous, G.: Water Reuse: From ancient to the modern times and future. Front. Environ. Sci. 6:26. doi:10.3389/fenvs.2018.00026, 2018.

[16] Bulu A. Historical development of hydrology. BALWOIS, Ohrid, North Macedonia. 2010;**9**

[17] Shamasastry R. Translation of 'Arthasastra' by Kautilya. Gov. Oriental Library Series, Bibliotheca Sanskrita, No. 37, Part 2. Bangalore; 1915

[18] Danby H. Translation of 'the Mishnah'. Oxford, UK: Oxford University Press; 1933

[19] Biswas, A. K.: A Short History of Hydrology, The Progress of Hydrology, Proceedings of the First International Seminar for Hydrology Professors, Volum II., 914-934, University of Illinois, Illinois, USA, 1969.

[20] Hollar S. Inventors and Innovations: Pioneers in the World of Weather and Climatology. New York: Britannica Educational Publishing in Association with Rosen Educational Services; 2013

[21] Koutsoyiannis, D.: Water control in the Greek cities. Water systems and urbanization in Africa and beyond Uppsala, Univ. of Uppsala, Sweden, 1-2 March 2012, 2012.

[22] Koutsoyiannis, D. and Angelakis, A. N.: Hydrologic and Hydraulic Sciences and Technologies in Ancient Greek Times. The Encycl. of Water Sci., Markel Dekker Inc., (B.A. Stewart and T. Howell, Eds.), Madison Ave. New York, N.Y., USA, 415-417, 2003.

[23] Koutsoyiannis D, Mamassis N, Tegos A. Logical and illogical exegeses of hydrometeorological phenomena in ancient Greece. Water Science and Technology: Water Supply. 2007;**7**(1):13-22. DOI: 10.2166/ws.2007.002

[24] Webster EW. Meteorologica. In: Ross WD, editor. The Works of Aristotle, Vol. 3. Oxford: Clarendon Press; 1955

[25] Horstmeyer, S. L.: An Outline of the History of Meteorology, available online: http://www.shorstmeyer.com/msj/geo165., 56, 2005.

[26] The Irish Times: Pliny the Elder: Sailor and Meteorologist. https://www.irishtimes.com/news/pliny-the-elder-sailor-and-meteorologist-1.219939 (assessed 3/28/2020), 2020.

[27] Papadopoulos E. Heron of Alexandria (c. 10-85 AD). Athens, Greece: National Technical Univ. of Athens; 2007

[28] Koutsoyiannis, D. and Patrikiou, A.: Water control in Ancient Greek cities. In: Water and Urbanization, T. Ostigard, (ed.). 130-148, 2013.

[29] Strangeways I. Improving precipitation measurements. International Journal of Climatology. 2004;**24**(11):1443-1460. DOI: 10.1002/joc.1075

[30] Strangeways I. A History of Raingauges. TerraData Ltd., http://www.rmets.org/sites/default/files/pdf/presentation/20100417-strangeways.pdf, (accessed 10 May 2015); 2010

[31] Hameen-Anttila J. The Last Pagans of Iraq: Ibn Wae Shiyya and his Nabatean Agriculture. Leiden, The Netherlands: Koninklijke Brill NV; 2006

[32] Hirschberger J. A Short History of Western Philosophy. Boulder, CO: Westview Press; 1977

[33] Reynolds R, Hammond JC, Smith F, empest S. In: Gilbert R, John D, Colson RS, et al., editors. Eyewitness Companions: Weather-Forecasting, Weather Phenomena, Climate Change, Meteorology. New York: Dorling Kindersley Publishing; 2008

[34] Gallica: Les Meteores. https://gallica.bnf.fr/ark:/12148/btv1b86069594/f243.item (assessed 3/28/2020), 2020.

[35] Bernal A. Q: Discover ideas about Weather Predictions. https://www.pinterest.es/pin/495325659001964022/ (assessed 3/5/2020), 2020.

[36] Ferrari, J.: Un breve cenno alla storia. http://pcfarina.eng.unipr.it/DispenseArch00/ferrari/ferrari.htm (assessed 3/5/2020), 2000.

[37] Francesco Folli's Hygrometer: http://himetop.wikidot.com/francesco-folli-s-hygrometer (assessed 6/15/2020), 2020.

[38] Weissman SA. Temperatures near the absolute zero. American Journal of Physics. 1947;**15**(6):451-457

[39] Piana ME. Hadley Cells. Harvard Press. https://www.seas.harvard.edu/climate/eli/research/equable/hadley.html (assessed 3/28/2020); 2020

[40] Darrigol O, Frisch U. From Newton's mechanics to Euler's equations. Physica D: Nonlinear Phenomena. 2008;**237**(14-17):1855-1869. DOI: 10.1016/j.physd.2007.08.003

[41] Allaby M. Dangerous Weather: A Chronology of Weather. Inc, New York, USA: Facts on File; 1998

[42] Biswas, A. K.: The Automatic Rain-Gauge of Sir Christopher Wren, F.R.S., Notes and Records of the Royal Society of London, 22 (1/2), Royal Society, 94-104, 1967.

[43] Thayer, B.: Theophrastus. De Signis by Theophrastus. Published in Vol. II of the Loeb Classical Library Edition of the Enquiry into Plants, 1926, http://penelope.uchicago.edu/Thayer/E/Roman/Texts/Theophrastus/De_signis*.html, (accessed 4 April 2015), 2012.

[44] Walker G. An Ocean of Air: A Natural History of the Atmosphere. London, UK: Bloomsbury; 2010

[45] Elsin, H.: Nilometer. In: Encyclopaedia of the History of Science, Technology, and Medicine in Non-Western Cultures (H. Selin, Ed.). Springer Netherlands, Absterdam, The Netherlands, p. 1753, 2008.

[46] The Editors of Encyclopaedia Britannica. John Frederic Daniell: British Chemist. Encyclopaedia Britannica, www.britannica.com/EBchecked/topic/151016/John-Frederic-Daniell, (accessed 10 May 2015); 2015

[47] Rainfall B. On the distribution of rain in space and time over the British Isles during the year 1908 as recorded by more than 4500 observers in Great Britain and Ireland, and discussed with articles upon various branches of Rainfall work. Nature. 1909;**81**:514-515. DOI: 10.1038/081514b0

[48] Pedgley DE. A short history of the British Rainfall organisation. Journal of Meteorology. 2003;**28**(276):74-75. DOI: 10.1002/wea.579

[49] Dalezios NR. Evolution of meteorology: From aristotle to satellites, proceedings of the 1st IWA regional symposium on water. Wastewater and Environment, Patras, Greece. 2014;**201-211**

[50] Dalezios NR, Nastos P. Milestones of the diachronic evolution of meteorology. Intern. J. Global Environ. Issues. 2016;**15**(1/2):49-69. DOI: 10.1504/IJGENVI.2016.074363

[51] Monmonier M. Air Apparent: How Meteorologists Learned to Map, Predict, and Dramatize Weather. Chicago: University of Chicago Press; 1999

[52] Walker M. History of the Meteorological Office. New York: Cambridge University Press; 2012

[53] Brassart, F.: Meteorograph by Secchi. Museum of INAF-Rome Astronomical Observatory, Monte Porzio Catone. Roma. https://tracieloeterra.bicentenarioangelosecchi.it/glistrumenti/dry-meteorograph/?lang=en (assessed 3/5/2020), 1867.

[54] Atlas Abscura: Meteorological Pillar of the 19th Century. https://www.atlasobscura.com/places/meteorological-pillar-of-the-19th-century (assessed 3/8/2020), 2020.

[55] History of the National Weather Service. Public Affairs Office, NWS and NOAA, http://www.nws.noaa.gov/pa/history/, (accessed 1 May 2015); 2012

[56] DuBois J, Multhauf R, Ziegler C. The Invention and Development of the Radiosonde, with a Catalog of Upper-Atmospheric Telemetering Probes in the National Museum of American History: Smithsonian Institution. Washington DC: Smithsonian Institution Press; 2002

[57] Jewell R. The Bergen School of Meteorology: The cradle of modern weather-forecasting. Bull. Am. Meteorol. Soci. 1981;**62**(6):824-830. DOI: 10.1007/BF00878151

[58] Donn ML. Meteorology. 4th ed. New York: McGraw-Hill Book Company; 1975

[59] Lee HDP. Aristotle: Meteorologica, Loeb Classical Library No. 397. Cambridge and London: Harvard University press and Heinemann; 1952

[60] Hugo Hildebrand Hildebrandsson. Encyclopedia Britannica 1911. ITA, www.theodora.com/encyclopedia/h2/hugo_hildebrand_hildebrandsson.html, (accessed 27 April 2015); 2011

[61] Overview of the Met Office: (Updated 2012) Met Office, Met Office, http://www.metoffice.gov.uk/news/in-depth/overview, accessed 15 September 2015, 2015.

[62] Barnett C. Rain: A Natural and Cultural History. New York: Crown Publishers; 2015

[63] Supercomputers: Met Office, http://www.metoffice.gov.uk/news/in-depth/supercomputers, (accessed 2 July 2015), 2014.

[64] Henson R. Weather on the Air: A History of Broadcast Meteorology. Boston, USA: American Meteorological Society; 2010

[65] Cobb A. Weather Observation Satellites. New York: The Rosen Publishing Group, Inc.; 2003

[66] Stunder B. What Is the Difference between the AVN, MRF, and GFS Models? Air Resources Laboratory, NOAA, https://www.ready.noaa.gov/faq_md15.php, (accessed 11 September 2015); 2010

[67] AghaKouchak A, Farahmand A, Melton FS, Teixeira J, Anderson MC, Wardlow BD, et al. Remote sensing of drought: Progress, challenges and opportunities. Reviews of Geophysics. 2015;**53**(2):452-480. DOI: 10.1002/2014RG000456

[68] Sun Q, Miao C, Duan Q, Ashouri H, Sorooshian S, Hsu KL. A review of global precipitation data sets: Data sources, estimation, and intercomparisons. Reviews of Geophysics. 2018;**56**(1):79-107. DOI: 10.1002/2017RG000574

[69] Bruintjes, R., and Terblanche, D.: Report of the Expert Team on Weather Modification Meeting, WMO and ETMW, Phitsanulok, Thailand, 17-19 March 2015, 4-6, 2015.

[70] Levine, J.: NASA's Ikhana Could Fly in the NAS Unaccompanied in 2018. https://www.nasa.gov/centers/armstrong/feature/Ikhana_could_fly_unaccompanied_in_2018.html (assessed 3/8/2020), 2018.

[71] Kumar R, Bahuguna IM, Ali SN, Singh R. Lake inventory and evolution of glacial lakes in the Nubra-Shyok basin of Karakoram range. Earth Sys. Environ. 2020;**4**(1):57-70. DOI: 10.1007/s41748-019-00129-6

[72] Saleem A, Dewan A, Rahman MM, Nawfee SM, Karim R, Lu XX. Spatial and temporal variations of erosion and accretion: A case of a large Tropical River. Earth Sys. Environ. 2020;**4**(1):167-181. DOI: 10.1007/s41748-019-00143-8

[73] Goodell J. How to Cool the Planet: Geoengineering and the Audacious Quest to Fix Earth's Climate. Boston, Massachusetts and New York, New York: Houghton Mifflin Harcourt; 2010

[74] Wigington, D.: Global Weather Modification Assault Causing Climate Chaos and Environmental Catastrophe, Geoengineering Watch, http://www.geoengineeringwatch.org/global-weathermodification-assault-causing-climate-chaosand-environmental-catastrophe-2/, (accessed 18 March 2016), 2014.

[75] Dale, M., Davies, P. and Harrison, T.: February. Review of recent advances in UK operational hydrometeorology, Proceed. Institution Civil

Engineers-Water Manage. 165, 2, 55-64, doi:10.1680/wama.2012.165.2.55, 2012.

[76] Sene K. Hydrometeorology-Forecasting and Applications. New York: Springer; 2010

[77] Zanchetta AD, Coulibaly P. Recent advances in real-time pluvial flash flood forecasting. Water. 2020;**12**(2):570. DOI: 10.3390/w12020570

[78] Fang MC, Tsai KY, Fang CC. The effect of hydrometeorology on ship collision avoidance in heavy traffic areas with a simplified simulation model. J. Marine Sci. Technol. 2019;**27**(3):235-245. DOI: 10.6119/JMST.201906_27(3).0006

[79] Ntelekos AA, Smith JA, Baeck ML, Krajewski WF, Miller AJ, Goska R. Extreme hydrometeorological events and the urban environment: Dissecting the 7 July 2004 thunderstorm over the Baltimore MD metropolitan region. Water Resources Research. 2008;**44**(8). DOI: 10.1029/2007WR006346

[80] Sawada Y, Nakaegawa T, Miyoshi T. Hydrometeorology as an inversion problem: Can river discharge observations improve the atmosphere by ensemble data assimilation? Journal of Geophysical Research – Atmospheres. 2018;**123**(2):848-860. DOI: 10.1002/2017JD027531

[81] Yang J, Wang ZH, Georgescu M, Chen F, Tewari M. Assessing the impact of enhanced hydrological processes on urban hydrometeorology with application to two cities in contrasting climates. Journal of Hydrometeorology. 2016;**17**(4):1031-1047. DOI: 10.1175/JHM-D-15-0112.1

[82] Crow WT, Huffman GF, Bindlish R, Jackson TJ. Improving satellite-based rainfall accumulation estimates using spaceborne soil moisture retrievals. Journal of Hydrometeorology. 2009;**10**:199-212. DOI: 10.1175/2008JHM986

[83] Herrnegger M, Nachtnebel HP, Schulz K. From runoff to rainfall: Inverse rainfall-runoff modeling in a high temporal resolution. Hydrol. Earth Sys. Sci. 2015;**19**:4619-4639. DOI: 10.5194/hess-19-4619-2015

[84] Vrugt JA, ter Braak CJF, Clark MP, Hyman JM, Robinson BA. Treatment of input uncertainty in hydrologic modeling: Doing hydrology backward with Markov chain Monte Carlo simulation. Water Resources Research. 2008;**44**:W00B09. DOI: 10.1029/2007WR006720

[85] Aubert D, Loumagne C, Oudin L. Sequential assimilation of soil moisture and streamflow data in a conceptual rainfall-runoff model. Journal of Hydrology. 2003;**280**(1-4):145-161. DOI: 10.1016/S0022-1694(03)00229-4

[86] Brocca L, Ciabatta L, Massari C, Moramarco T, Hahn S, Hasenauer S, et al. Soil as a natural rain gauge: Estimating global rainfall from satellite soil moisture data. J. Geophysic. Res. Atmos. 2014;**119**:5128-5141. DOI: 10.1002/2014JD021489

[87] Brocca L, Moramarco T, Melone F, Wagner W. A new method for rainfall estimation through soil moisture observations. Geophysic. Res. Let. 2013;**40**:853-858. DOI: 10.1002/grl.50173

[88] Brocca L, Moramarco T, Melone F, Wagner W, Hasenauer S, Hahn S. Assimilation of surface-and root-zone ASCAT soil moisture products into rainfall-runoff modelling. IEEE Transact. Geosci. Remote Sens. 2011;**50**(7):2542-2555. DOI: 10.1109/TGRS.2011.2177468

[89] Crow WT, van den Berg MJ, Huffman GJ, Pellarin T. Correcting rainfall using satellite-based surface soil moisture retrievals: The soil moisture analysis Rainfall tool (SMART). Water Resources Research. 2011;**47**:W08521. DOI: 10.1029/2011WR010576

[90] Koster RD, Brocca L, Crow WT, Burgin MS, De Lannoy GJM. Precipitation estimation using L-band and C-band soil moisture retrievals. Water Resources Research. 2016;**52**:7213-7225. DOI: 10.1002/2016WR019024

[91] Mohammed, K., Leconte, R. and Trudel, M.: January. Data Assimilation of Remotely Sensed Soil Moisture in Hydrological Modeling to Improve Flood Forecasting. In 100th American Meteorological Society Annual Meeting. AMS, 2020.

[92] Wu DS, Wang WJ, Yu SB, Zhou SH, Zhang J. Study on relationship between red tide of year 2002 in Guangdong coastal water and the marine hydro-meteorology. J. Tropic. Meteorol. 2011;**27**:271-277

[93] UNISDR (United Nation Office for Disaster Risk Reduction). The Human Cost of Weather-Related Disasters 1995-2015. The Centre for Research on the Epidemiology of Disasters (CRED), United Nation Office for Disaster Risk Reduction (UNISDR). Available via DIALOG. http://www.unisdr.org/2015/docs/climatechange/COP21_WeatherDisastersReport_2015_FINAL.pdf. (Accessed 19 Sept 2020); 2015

[94] Mao, H., Paul, O. K., Yang, N. and Li, L.: Smart Arduino Sensor Integrated Drone for Weather Indices: Prototype. In Drones-Applications. IntechOpen. Drones: Applications (Editor: George Dekoulis). Electronic book. ISBN: 178923284, 2018.

[95] Monmonier M. Air Apparent: How Meteorologists Learned to Map, Predict, and Dramatize Weather. Chicago, USA: University of Chicago Press; 2000

[96] Moore RR, Collins MD, Moore RR, Collins MD. Personal severe weather warning microchip and pressure sensor. U.S. Patent. 2006;7:066,020

[97] Vas Á, Fazekas Á, Lehotai B, Nagy G, Tóth L. Microcontroller-based network for meteorological sensing and weather forecast calculations. Carpathian Journal of Electronic and Computer Engineering. 2012;**5**:139

[98] Willis PA, Creamer JS, Mora MF. Implementation of microchip electrophoresis instrumentation for future spaceflight missions. Analytic. Bioanalytic. Chem. 2015;**407**(23):6939-6963. DOI: 10.1007/s00216-015-8903-z

[99] Zheng S, Wang G, Huang X, Miao C, Xing W, Chen S, et al. Improvement and Design of Transmitter Modifier Wind Cooling Protection for CINRAD/CB weather radar. J. Geosci. Environ. Protect. 2018;**6**(11):139-146. DOI: 10.4236/gep.2018.611011

[100] Almazroui, M., Saeed, S., Saeed, F., Islam, M. N. and Ismail, M.: Projections of Precipitation and Temperature over the South Asian Countries in CMIP6. Earth Sys. Environ., 4(2), 297-320, doi:10.1007/s4174 8-020-00157-7, 2020a.

[101] Almazroui, M., Saeed, F., Saeed, S., Islam, M. N., Ismail, M., Klutse, N. A. B. and Siddiqui, M. H.: Projected Change in Temperature and Precipitation Over Africa from CMIP6. Earth Sys. Environ., 1-21, doi:10.1007/s00704-015-1715-4, 2020b.

[102] Anandharajan, T. R. V., Hariharan, G. A., Vignajeth, K. K. and Jijendiran, R.: Weather monitoring using artificial intelligence. In 2016 2nd international conference on computational intelligence and networks (CINE) (106-111). IEEE, 2016.

[103] Liu, J. N., Hu, Y., He, Y., Chan, P. W. and Lai, L.: Deep neural network modeling for big data weather forecasting. In Information Granularity, Big Data, and Computational Intelligence (389-408). Springer, Cham, 2015.

[104] McGovern A, Elmore KL, Gagne DJ, Haupt SE, Karstens CD, Lagerquist R, et al. Using artificial

intelligence to improve real-time decision-making for high-impact weather. Bull. Am. Meteorol. Soci. 2017;**98**(10):2073-2090. DOI: 10.1175/BAMS-D-16-0123.1

[105] Carranco JS, Salgado FD, Sellers C, Torres H. Comparative Analysis of Meteorological Monitoring Using an Integrated Low-Cost Environmental Unit Based on the Internet of Things (IoT) with an Automatic Meteorological Station (AWS), In 2017 IEEE Second Ecuador Technical Chapters Meeting (ETCM) (1-6). IEEE; 2017

[106] Kavitha, N. and Madhumathy, P.: Development of an IOT-Based Atmospheric Fine Dust Monitoring System. In Internet of Things, Smart Computing and Technology: A Roadmap Ahead (263-279). Springer, Cham, 2020.

[107] Onal AC, Sezer OB, Ozbayoglu M, Dogdu E. December. Weather data analysis and sensor fault detection using an extended IoT framework with semantics, big data, and machine learning. In: 2017 IEEE International Conference on Big Data (Big Data) (2037-2046). IEEE. 2017

[108] Pradeep, S. and Sharma, Y. K.: Storing live sensor data to the platforms of internet of things (IoT) using Arduino and associated microchips. In Proceedings of the Third International Conference on Computational Intelligence and Informatics (1-15). Springer, Singapore, 2020.

Chapter 2

Perspectives of Hydrologic Modeling in Agricultural Research

Miha Curk and Matjaž Glavan

Abstract

For decades agricultural research was done in the field or laboratories, but with the rise of computer science, hydrologic modeling became another essential tool for environmental impact studies. Many types of models can be used, each with its strengths and weaknesses in terms of accuracy, speed, and amount of input data needed. Models can be used on different scales and simulate very different processes. Based on a literature review, APEX (Agricultural Policy Extender) and SWAT (Soil and Water Assessment Tool) models are the most popular for environmental research in agronomy. An important share of modeling work in agronomic studies is focused on pollution research, mainly nutrient and pesticide leaching and soil erosion processes. Other topics include simulating the effects of irrigation and other agricultural practices and studying the impact of extreme weather events and climate change. When working with model results, it is crucial to be mindful of inevitable uncertainties and consider them during interpretation. Modeling is gaining importance in agronomic research in Slovenia, with many studies done in the recent decade and more underway.

Keywords: hydrologic modeling, agriculture, agronomy, model applications, agricultural pollution mitigation, model uncertainty

1. Introduction

For decades, agricultural research was predominantly done in the field or laboratories. Such in situ research results are usually exact, but experiments are time-consuming. Due to different natural conditions, their validity is usually limited to the small area under study. For example, a study on crop yield in a specific area and under a specific agricultural management only applies to such conditions, and for different conditions, a new experiment needs to be devised. Even when results are visible and relatively easy to measure (like crop yield), the spatial differences prevent us from extrapolating them over a large area without some degree of error. This error only gets larger when research goals include measuring more complicated phenomena like pollution.

With strict limitations in environmental policy (Water Framework Directive (WFD) (Directive 2000/60 EC) in EU; Clean Water Act in the USA), an important branch of agricultural research is focused on mitigating pollution from agricultural activities. Nitrate leaching, sediment erosion, ammonia emissions, or pesticide pollution are hard to measure reliably over larger areas. Natural processes like plant growth also take time, making it hard to conduct conventional field trials to study alternative fertilization or pesticide application methods. There is another way to estimate the impact of different practices, which involves modeling. Modeling did not render the field studies useless because their results are indispensable as input and validation data.

Several decades ago, with the rise of computer science, different mathematical model approaches were developed to simulate parts of the natural system. Hydrologic modeling in the 1970s quickly became a trendy way of studying the physical processes behind water and nutrient cycling in soil, plants, and whole ecosystems. By coupling several of the more focused models (plant growth, nutrient and water cycle, etc.), more complex models were developed, enabling fast (relative to in situ research) estimations of outcomes of different climate scenarios, land-use changes, etc.

Hydrologic models are not intuitively connected to agriculture, as their use is far more prevalent in other fields. An article on 'Brief history of agricultural systems modeling' by Jones et al. [1], for example, does not even mention them. Despite that, hydrologic modeling is an essential tool in an increasingly important field of agricultural research, the environmental impact studies. The use of hydrologic models enabled fast advances in understanding pollutant movement in different ecosystems, making pollution mitigation strategies easier to evaluate.

This chapter will discuss different hydrologic models used in agricultural research, their strengths and weaknesses, their potential for agricultural pollution mitigation, and the uncertainty associated with model results. Lastly, we will look into practical applications and present some case studies from Slovenia.

2. Hydrologic models – an overview

Science and research in some fields depend strongly on modeling these days, and the world would probably be different if this tool were not available to us. Hundreds of models were developed over several decades, some simple and some very complex. Each model usually has its particular purpose, though some are quite elaborate and enable the user to model several extensive systems processes simultaneously. In an excellent paper about the evolution of hydrologic models, Clark et al. [2] discussed the challenges of designing hydrologic models that are as close to physical realism as possible while still keeping them simple enough and practical. The authors summarized that there were many noteworthy advances in their development in the last years, as improvements in representations of hydrologic processes by mathematical functions, parameter estimation, and optimizing computing resources by justifiable model simplifications. Some of the main goals for the future they mention are improvements of the basic hydrologic processes understanding, of parallel processing, of cooperation between different model developers in order to find the best methods, of the model analysis methods in order to minimize uncertainty, but also enhancement of the developer-field scientist interaction to promote usability and most importantly improvement and clarification of the construction of the models themselves, to enable more specific add-ons and better modularity.

Hydrologic models are divided into several different categories, depending on how they are structured and represent spatial processes [3]. Based on the structure, models are divided into empirical, conceptual, and physical; based on spatial distribution into lumped, semi-distributed, and distributed. Hydrologic models used in agriculture are almost exclusively either conceptual or physical and semi-distributed or distributed. Empirical and lumped models are not practical for such applications because the former models are very exact and depend heavily on large amounts of measured input data, and the latter disregard the spatial variability inside the modeled area. The difference between conceptual and physical models is that conceptual models consist of simplified equations representing water storage in the catchment, and physical models are based on physical laws and equations based on measured hydrologic responses.

Consequently, the latter are more difficult to calibrate and require many parameters, but the former rarely consider spatial variability within the catchment and are better to use in large catchments with limited data and computational times. On the other hand, distributed models are the ones where the modeled area is divided into smaller cells by a grid of specific size, and semi-distributed ones divide it into specific shapes that represent essential features inside the area. Consequently, the former models are data-intense with long computational times, but the latter risk loss of spatial resolution as the sub-catchments get larger [3].

As mentioned before, there are plenty of models a researcher can consider for his work. Malone et al. [4] discussed the parameterization guidelines and considerations and mentioned at least 15 different models. Google Scholar search was performed to assess the popularity of some of the mentioned models in the agricultural context, with a query: "model acronym" model AND (agronomy OR agriculture OR farm). The number of hits is written in brackets after each model acronym: Watershed Analysis Risk Management Framework (WARMF) (400), HYDRUS (8900), European Hydrological System Model (MIKE-SHE)(3900), DRAINMOD (2500), Soil and Water Assessment Tool (SWAT)(45,100), Environmental Policy Integrated Climate and Agricultural Policy/Environmental Extender (EPIC/APEX) (178,000/172,000), Root Zone Water Quality Model (RZWQM)(2000), Better Assessment Science Integrating Point and Nonpoint Sources (BASINS)(1000), Hydrological Simulation Program – FORTRAN (HSPF)(6000). The EPIC/APEX models are the most used in agricultural context with over 170,000 hits, followed by SWAT with 45,000 hits based on the search results. Other models achieved less than 10,000 hits and seemed far less popular. EPIC/APEX, SWAT, and MIKE-SHE models are presented in more detail in **Table 1**, based on a comparison study by Golmohammadi et al. [5].

According to the study [5], SWAT and MIKE-SHE were recognized as very well-performing models (in terms of river discharge), and SWAT was considered the better of the two when simulating processes in agricultural catchments. On the other hand, APEX was recognized as perfect for scenario assessment on farm scale, due to its many options in management practices (different irrigation types, drainage, buffer strips, terraces, fertilization management, etc.), but also economic

Model acronym	Structure	Spatial distribution	Strengths	Weaknesses
EPIC-APEX	Physical	Distributed	Suitable for simulation of many agricultural management scenarios	Only for small-scale watersheds or farms
SWAT	Physical	Semi-distributed	Shorter computational times due to lumping of spatial units in the form of hydrological response units (HRUs)	Simplified spatial distribution
MIKE-SHE	Physical	Distributed	Built-in graphics and post-processor for calibration and analysis of results	Long computational times, data-intensive

Table 1.
Comparison of three models commonly used in agricultural research with model type, strengths, and weaknesses.

analysis of measures. Management practices options are also available in SWAT, but it is usually considered better for larger scale watersheds. However, the authors conclude that no single model is superior under all conditions and that the model's performances are very site-specific. In light of agricultural research, the EPIC/APEX model is most useful in small catchments with lots of known data. MIKE-SHE is most useful for large areas when computational time is not a constraint, and data is plentiful. SWAT seems to be somewhere in between – allowing short computational times even in large watersheds but enabling the reasonably accurate agricultural management simulation. Another advantage of SWAT is its modularity – it is easy to link it to other more specific models.

Therefore, models are the most appropriate and cost-effective method for assessing different agricultural management strategies and their impact on the environment. Outputs can be calculated daily, monthly, and yearly and can be used to study the long-term effects of climate change adaptation and short term influence on crop yields, state of the soil, etc.

3. Agricultural pollution research – a challenge

This chapter will dive further into the important question: What circumstances make the models more suited for agricultural impact studies than field trials? As discussed in the introduction, environmental research is an essential field of science today, and water pollution is the part where hydrologic models can be of great help. Water quality is one of the Sustainable Development Goals in the 2030 Agenda for Sustainable Development, and an FAO report on the topic [6] lists agriculture as one of the three major sources of water pollution, along with human settlements and industry. Main water pollution threats posed by agriculture and some possible mitigation strategies are presented in **Table 2**.

Different environments are very diverse, and natural conditions are spatially specific. The dynamics of processes leading to pollution can differ from place to place, even if they are not far apart. Testing the impacts of different mitigation

Threat	Main sources	Threatened waterbody	Possible mitigation strategies
Nitrate	Fertilization, manure storage	Surface and groundwater	Balanced fertilization, manure storage in contained areas, managed grazing, catch crops
Phosphorus	Fertilization, soil erosion	Surface water	Balanced fertilization, reducing soil erosion
Sediment	Soil erosion	Surface water	Cover crops, reduced tillage, tilling parallel to contour lines, terracing, managed grazing
Pesticides	Plant protection	Surface and groundwater	Planting hardy or resistant varieties, increasing biodiversity, balanced application, proper waste disposal
Veterinary medicines	Livestock healthcare	Surface and groundwater	Use according to international guidelines
Salinization	Irrigation	Surface water	Minimize drainage, use less water demanding crops

Table 2.
Main pollutants from agriculture, their sources, threatened water bodies, and theoretical mitigation strategies.

strategies on a large scale would be very slow and expensive if the only tool we had were field trials. Luckily, the models provide an alternative. An FAO report [6] describes it very clearly: "Models provide ... holistic understanding of problems by identifying relationships (cause and effect), and future predictions (scenarios). Models can simulate the fate of pollutants and the resulting change in the state of water quality and help understand the impacts on human health and ecosystems. Models can also help in determining the effectiveness and costs of remedial actions."

Models are not only useful in predicting the efficiency of potential mitigation measures; they are just as often used as a tool to help understand the current state of pollution. Except in some cases (i.e., soil erosion), most pollutant threats are generally invisible to human eyes, and the only way to understand the extent of pollution is often (besides monitoring by point measurements) through the model simulations. Of course, this does not mean that field monitoring is outdated. Measurements can provide important starting points and input or calibration data for models to perform accurately. Let us put it this way: Monitoring is a way to detect that a natural system is threatened, field trials allow us to obtain important data about processes on a local scale, and modeling is a tool that helps us understand the extent of pollution in a broader scale and provide information on promising mitigation strategies.

Which model to use depends strongly on the scope of research one intends to conduct. Apart from the already discussed division by structure or spatial distribution, models can also be divided into groups based on the expected outcomes. Field-scale models are generally only capable of simulating local processes, like plant growth with water and solutes movement through soil, but do that quickly and quite reliably because little input data (like soil type, weather, cropping system etc.) is needed, and most of it can be measured in the field. No particular skill is usually required to set up such a model. As models transition into regional or catchment scale, more and more of the input data is interpolated across larger areas or not known precisely. In such cases, merely inputting the available measured data will not result in a well-performing model because gaps in the so-called "hard data" (the measurements) are usually too big. The modeler needs to find a way to fill those gaps with "soft data" – possible data ranges characteristic to specific conditions in the area. Ways to learn about soft data are technical field trips, consulting local experts, examining data from similar areas elsewhere etc.

Besides the already discussed pollution studies, hydrologic models can also be used in other agronomic research branches. They can be set up to analyze impacts of droughts, other weather events, and even climate change, simulate effects of irrigation or drainage, study the water balance of different crops, model water retention capabilities of soils, etc. [7–9].

4. Model uncertainty – why models can be misleading

As discussed in the previous chapter, model simulations are a blessing to researchers, but they can be misleading. If the models are used negligently or if the results are misinterpreted, they can very well be a curse, providing us with dubious information. No model is entirely accurate, and even the most experienced modelers in the world do not claim their model results are 100% certain. Quite the opposite experienced modelers will know very well what their setups' flaws and uncertainties are. It is often said that modeling is an art as much as a science because the modeler needs to balance process resolution, computational speed, and accuracy to ensure a reasonable output. Furthermore, he or she needs to overcome the challenge of presenting an enormous amount of information in a way that can be used to increase understanding of the system [10].

So how does one ensure that his model performs well enough? Several operations optimize the performance and improve our understanding of uncertainties: parameterization, sensitivity analysis, calibration, validation, and uncertainty analysis. Parameterization is the process of assigning data to model parameters. Theoretically, all the input data would be measured, but there are obstacles to that – firstly, not everything can be measured, and secondly, even some measurements are not entirely realistic. Therefore, "hard" data is input first, followed by "soft" data to the best of our knowledge. An article by Malone et al. [4] discusses parameterization in more detail. Once input data is inserted, sensitivity analysis is performed to find out what parameters are sensitive. If a parameter is sensitive, its changes significantly influence model results. If it is not, no matter how much we change it, the results will be similar. Sensitive parameters and those of which values we are uncertain are then modified during the process of calibration to match the model results as closely with observed values for river discharge, nutrient loads, crop yields, etc. Validation is executed next, possibly for different seasons, to ensure robustness and verify that the calibrated parameters results show good model performance outside of the calibration period. Moreover, uncertainty analysis shows us what the uncertainties in the model results are. Sensitivity and uncertainty analysis might seem like the same thing, and they are in a way, but the former points out how much different input parameters influence the final results, while the latter focuses on the uncertainty of final results directly [11].

With each model, there are several ways one might go about the abovementioned procedures. In the past, manual calibration was the norm, and it meant manually changing different parameter values until the desired matching of observed and simulated data was achieved. With large numbers of parameters in models, this method is time-consuming and requires quite some experience, and some authors suggest against using it [12] because it is hard to achieve a range of possible simulations in this way. This leads us to the next topics, which are automated calibration, sensitivity, and uncertainty tools. Many models have a built-in or standalone program developed specifically for them (MIKE-SHE has a built-in tool, SWAT-CUP [13] is a standalone tool for SWAT, etc.). There are also quite advanced but universal tools that can work with different models, like PEST: Model-Independent Parameter Estimation and Uncertainty Analysis [14].

For parameterization, it is essential to have good data. Any type of data is not equally useful in modeling work, and different types of data may be useful in different situations. Soil data, for example, can be carefully measured, or pedotranspher functions can be applied to calculate it. But which is better depends on what type of model is used. For a field-scale model, acquiring measured data is usually beneficial, and the scale of operation is also feasible. For larger-scale models, though, it depends on the accuracy of measurements, heterogeneity of soil in an area, and many other factors. Usually, large scale models require so many measurements that acquiring them is no longer feasible, and one must rely on data provided by different databases for the area. Interestingly, measured and calculated soil characteristics can vary quite a lot, as shown by our data in **Table 3**.

Differences in the presented data could result from soil cracks, earthworm burrows, agricultural management, and others, which were not accounted for in one of the methods. Conveniently, soil parameters are almost always calibrated because they influence the water cycle significantly. Based on previous modeling experience, we found that for large-scale models (especially since soil hydraulic properties measurements are expensive, time-consuming, and require special equipment), it is usually more than adequate to use pedotranspher calculations as a basis. From there,

		Hydraulic conductivity [mm/h]		Available water capacity [cm^3 water/cm^3 soil]	
Soil type	*Soil layer*	*Measured*	*Calculated*	*Measured*	*Calculated*
Calcaric Fluvisol	A1	6.0	14.0	0.14	0.14
	A	1600.8	14.2	0.14	0.14
	Bv	92.1	19.2	0.17	0.14
	I	316.9	24.4	0.10	0.12
	II	107.3	43.4	0.10	0.11
	III	417.3	30.8	0.10	0.07
Dystric Cambisol	A	4257.9	9.5	0.14	0.14
	Ap	85.4	3.9	0.17	0.14
	Bv	1301.0	2.0	0.17	0.12
Calcaric Fluvisol	Ap	160.4	9.6	0.14	0.16
	AB	160.4	10.0	0.14	0.16
	Bv	163.3	9.9	0.17	0.16
	Bg	169.2	9.7	0.10	0.16

Table 3.
Presentation of soil hydraulic properties in cases of measurement and calculation.

the most realistic values can be determined during calibration, thus not modifying the "expensive" measured soil data.

Another note concerns the calibration data. It is vital to choose the data that represents a prevailing hydrological process in the catchment. For example, suppose discharge is altered too much by human activity or other processes not accounted for in the model structure, or point sources in the watershed contribute a significant share of the water into the cycle. In that case, it might be better to use an alternative dataset, although most other model applications used discharge data for basic calibration. Besides discharge, soil moisture measurements (both satellite and in situ data) are gaining significance in the last years [15–17] and can be a useful alternative in areas where discharge data is not convenient or possible to use.

Model calibration and uncertainty analysis are a vast field of study, so we will not detail them here. There are several comprehensive papers and manuals on the topic [12, 14, 18, 19], and before diving into the calibration of a model, it is crucial to get as much knowledge on the topic as possible.

5. Model applications – recent case studies in Slovenia

Slovenia, as a European Union member state, had transposed the WFD to state law in 2002, and since then, much work was done in the field of environmental studies in agricultural areas. Slovenia's biggest issue regarding water protection is groundwater bodies under large river plains with relatively shallow soil profiles. While being very appropriate for agricultural and urban activities, they are also very vulnerable to nitrate and pesticide leaching. The state of water bodies is mostly good, except for some aquifers in the Northeast part of the country, where it seems groundwater recharge is not as strong due to less precipitation in the area. More details on water protection measures and laws in Slovenia can be found in [20].

Reviewing different modeling efforts in Slovenian agricultural areas is an excellent way to get insight into implementing hydrologic modeling in general. For this chapter, another Google Scholar search was conducted, this time with a query: (hydrologic OR water) AND model AND (agronomy OR agriculture OR farm) AND "Slovenia". The search was repeated in Slovene to find more studies that were not published in English. After a scan through the results, several interesting studies were selected, joined by some others we have known from previous work, and were for some reason not included in the search. Selected publications all fit into the category of hydrologic modeling in agricultural areas. In terms of scale, some of them feature large scale modeling of the whole country, some catchment scale, and another field-scale modeling. In terms of the type of model used, there are several of them, but SWAT model applications are the most frequent. Topics range from nitrate leaching and concentration in groundwater to sediment, phosphorus, and nitrate loads in surface waters, and even to weather extremes modeling, including droughts and climate change.

The whole country modeling effort to determine nitrogen reduction levels necessary to reach groundwater quality targets was a program led by Slovenian Environment Agency [21]. Hydrological model GROWA–DENUZ was coupled with agricultural N balances to simulate nitrate leaching for the whole country. Results indicate that stricter measures in vulnerable areas are crucial to meeting WFD thresholds, while additional state-wide measures are not necessary.

Several studies [22–24] were conducted in vulnerable areas where groundwater is not a good state. While studying nitrate leaching, just like the work above, they were limited to catchment scale, and the model used was SWAT. Several agricultural management scenarios were simulated to determine what type of management is the most effective at reducing nitrate leaching. Among many other findings, an important message is that careful placing of local measures based on soil characteristics can be just as effective at reducing nitrate leaching as applying more general limitations on a broader scale while allowing a much healthier socio-economic development agricultural sector.

One study [25] dealt with simulating the effect of different historical land-use scenarios on surface water quality. The SWAT model was used to determine how the land use documented on historical maps (18th, 19th, 20th' and 21st centuries) would impact river quality. Interestingly, the authors found that historical land-use patterns generally caused more erosion than the present, but even the present one is not the best for water organisms.

Another study [26] evaluated the effects of deforestation and increasing vineyard land use on surface water quality with the APEX model. Results show that though pollution increases with deforestation, proper protective measures (like vegetative buffer strips) can limit its scope.

In one case [27], a new model was developed based on equations from existing ones to simulate the effects of wastewater treatment implementation in an agricultural catchment. Results suggest that applying the measure of wastewater treatment did reduce nitrogen concentrations in the stream and increase phosphorus concentrations, which could worsen the situation in that specific catchment.

Finally, there were two studies [28, 29] dealing with controlling erosion and nutrient leaching in catchments with accumulation lakes.

Most of the described case studies took advantage of modeling to gain insight into differences between several agricultural management scenarios, which would be much more expensive and time-consuming if done with field trials. Interestingly, several studies also included some fieldwork, partially for input data acquisition, but mostly to collect reliable validation data like crop yields, nitrate concentration, soil properties, soil water showing that the "old" ways are still very viable. The best results can only be acquired if we employ the power of modeling and fieldwork combined.

6. Conclusions

In this chapter, we have discussed the perspectives of hydrologic modeling in agricultural research. The most frequently used hydrologic models were identified and reviewed in terms of their suitability for different applications in agronomy. A section evaluated the strengths and weaknesses of hydrologic models for agricultural research and highlighted potential applications. The importance of modeling in light of agricultural pollution mitigation was also be presented. Furthermore, the importance of input data quality and uncertainty analysis was discussed to highlight the potential risks associated with modeling. Examples of different case studies in Slovenia were referenced to review the recent agricultural modeling work in this country.

Future development in the field should concentrate on strengthening the interaction between model developers and users on one side and field scientists and farmers on the other, to make models more adept to specific practices and applications in different areas. This would strengthen the trust in modeling among agricultural scientists while expanding the recognition of modeling among the public and policymakers.

Overall, through this chapter and with every single one of the highlighted case studies, we hope to have strengthened the importance of hydrologic modeling in the agricultural sector. While model results cannot foretell the future, they can give us a useful range of possibilities to consider and discuss further despite their shortcomings and uncertainties. In conclusion, modeling has enabled important advances in agricultural hydrology studies and sped up research that would otherwise take much longer to conduct.

Author details

Miha Curk* and Matjaž Glavan
Biotechnical Faculty, University of Ljubljana, Jamnikarjeva 101, Ljubljana, Slovenia

*Address all correspondence to: miha.curk@bf.uni-lj.si

References

[1] Jones JW, Antle JM, Basso B, Boote KJ, Conant RT, Foster I, et al. Brief history of agricultural systems modeling. Agric Syst [Internet]. 2017 Jul 1 [cited 2020 Oct 21];155:240-54. Available from: http://creativecommons.org/licenses/by/4.0/

[2] Clark MP, Bierkens MFP, Samaniego L, Woods RA, Uijlenhoet R, Bennett KE, et al. The evolution of process-based hydrologic models: historical challenges and the collective quest for physical realism. Hydrol Earth Syst Sci [Internet]. 2017 [cited 2020 Oct 20];21:3427-40. Available from: https://doi.org/10.5194/hess-21-3427-2017

[3] Knightes C. An Overview of Rainfall-Runoff Model Types An Overview of Rainfall-Runoff Model Types. 2017;(September):0-29.

[4] Malone RW, Yagow G, Baffaut C, Gitau MW, Qi Z, Amatya DM, et al. Parameterization guidelines and considerations for hydrologic models. Trans ASABE. 2015;58(6):1681-703.

[5] Golmohammadi G, Prasher S, Madani A, Rudra R. Evaluating three hydrological distributed watershed models: MIKE-SHE, APEX, SWAT. Hydrology [Internet]. 2014;1(1):20-39. Available from: www.mdpi.com/journal/hydrologyArticle

[6] Fao & Iwmi. Water pollution from agriculture: a global review Executive summary. FAO IWMI [Internet]. 2017 [cited 2020 Oct 28];35. Available from: http://www.fao.org/3/a-i7754e.pdf

[7] Glavan M, Ceglar A, Pintar M. Assessing the impacts of climate change on water quantity and quality modelling in small Slovenian Mediterranean catchment - lesson for policy and decision makers. Hydrol Process. 2015;(29):3124-44.

[8] Glavan M, Kozinc Z, Pintar M. Kmetijstvo in Upravljanje S Poplavnimi Območji Ob Reki Dravi – Projekt Livedrava. In: Mišičev vodarski dan. 2015. p. 290-7.

[9] Urbanc J, Krivic J, Mali N, Ferjan Stanič T, Koroša A, Šram D, et al. Možnosti kmetovanja na vodovarstvenih območjih - končno poročilo [Internet]. Ljubljana; 2014 [cited 2018 Oct 29]. Available from: http://www.dlib.si/details/URN:NBN:SI:DOC-HWYIJSVV/?

[10] MacWilliams ML, Ateljevich ES, Monismith SG, Enright C. An overview of multi-dimensional models of the Sacramento-San Joaquin delta. San Fr Estuary Watershed Sci [Internet]. 2016 [cited 2020 Oct 28];14(4). Available from: https://doi.org/10.15447/sfews.2016v14iss4art2

[11] European Commission. What is difference between uncertainty and sensitivity analysis? [Internet]. EU Science Hub. 2015 [cited 2020 Oct 29]. Available from: https://ec.europa.eu/jrc/en/faq/what-difference-between-uncertainty-and-sensitivity-analysis-33469

[12] Abbaspour KC, Vaghefi S, Srinivasan R. A Guideline for Successful Calibration and Uncertainty Analysis for Soil and Water Assessment: A Review of Papers from the 2016 International SWAT Conference. Water [Internet]. 2017 Dec 22 [cited 2018 Mar 9];10(1):6. Available from: http://www.mdpi.com/2073-4441/10/1/6

[13] Karim C. Abbaspour. User Manual for SWAT-CUP, SWAT Calibration and Uncertainty Programs [Internet]. Duebendorf, Switzerland; 2015 [cited 2019 Dec 12]. Available from: https://swat.tamu.edu/media/114860/usermanual_swatcup.pdf

[14] Doherty J. Calibration and Uncertainty Analysis for Complex Environmental Models PEST: complete theory and what it means for modelling the real world [Internet]. 2015 [cited 2020 Oct 29]. Available from: www.pesthomepage.org

[15] Rajib MA, Merwade V. Improving soil moisture accounting and streamflow prediction in SWAT by incorporating a modified time-dependent Curve Number method. Hydrol Process [Internet]. 2016 Aug 15 [cited 2018 Nov 22];30(4):603-24. Available from: http://doi.wiley.com/10.1002/hyp.10639

[16] Kundu D, Vervoort RW, van Ogtrop FF. The value of remotely sensed surface soil moisture for model calibration using SWAT. Hydrol Process. 2017;31(15):2764-80.

[17] Curk M, Glavan M, Pintar M. Analysis of nitrate pollution pathways on a vulnerable agricultural plain in Slovenia: Taking the local approach to balance ecosystem services of food and water. Water (Switzerland) [Internet]. 2020 Mar 5 [cited 2020 Jul 15];12(3):1-15. Available from: https://www.mdpi.com/2073-4441/12/3/707

[18] D. N. Moriasi, J. G. Arnold, M. W. Van Liew, R. L. Bingner, R. D. Harmel, T. L. Veith. Model Evaluation Guidelines for Systematic Quantification of Accuracy in Watershed Simulations. Trans ASABE [Internet]. 2007 [cited 2018 Nov 21];50(3):885-900. Available from: http://citeseerx.ist.psu.edu/viewdoc/download?doi=10.1.1.532.2506&rep=rep1&type=pdf

[19] Abbaspour KC, Rouholahnejad E, Vaghefi S, Srinivasan R, Yang H, Kløve B. A continental-scale hydrology and water quality model for Europe: Calibration and uncertainty of a high-resolution large-scale SWAT model. J Hydrol [Internet]. 2015 May 1 [cited 2018 Feb 19];524:733-52. Available from: http://dx.doi.org/10.1016/j.jhydrol.2015.03.027

[20] Curk M, Glavan M. Groundwater Protection Legislation in Slovenia: Theory and Practice. In: Groundwater - Resource Characterisation and Management Aspects. IntechOpen; 2019.

[21] Andjelov M, Kunkel R, Uhan J, Wendland F. Determination of nitrogen reduction levels necessary to reach groundwater quality targets in Slovenia. J Environ Sci (China) [Internet]. 2014 [cited 2017 Nov 13];26(9):1806-17. Available from: https://ac.els-cdn.com/S1001074214000734/1-s2.0-S1001074214000734-main.pdf?_tid=1a0109de-c864-11e7-87d3-00000aab0f26&acdnat=1510571972_4b62008486fa3a5ee0999a47a3fa39e3

[22] Curk M, Glavan M, Pintar M. Analysis of nitrate pollution pathways on a vulnerable agricultural plain in Slovenia: Taking the local approach to balance ecosystem services of food and water. Water (Switzerland). 2020;12(3):1-15.

[23] Glavan M, Jamšek A, Pintar M. Modelling Impact of Adjusted Agricultural Practices on Nitrogen Leaching to Groundwater. H. Tutu, editor. Water Qual [Internet]. 2017;(January):113-33. Available from: http://www.intechopen.com/books/water-quality/modelling-impact-of-adjusted-agricultural-practices-on-nitrogen-leaching-to-groundwater

[24] Glavan M. Vpliv spremenjene rabe zemljišč na količino in kakovost vode v reki Reki v goriških brdih in reki Dragonji [Internet]. 2011. Available from: http://www.digitalna-knjiznica.bf.uni-lj.si/dd_glavan_matjaz.pdf

[25] Glavan M, Miličić V, Pintar M. Finding options to improve catchment

water quality-Lessons learned from historical land use situations in a Mediterranean catchment in Slovenia. Ecol Modell [Internet]. 2013 [cited 2017 Nov 10];261-262(262):58-73. Available from: https://ac.els-cdn.com/S0304380013002081/1-s2.0-S0304380013002081-main.pdf?_tid=fd88b0e2-c613-11e7-83f8-00000aacb35d&acdnat=1510317663_7e79081a15901f2755cd13c4a2a04aea

[26] Glavan M, Bele S, Curk M, Pintar M. Modelling Impacts of a Municipal Spatial Plan of Land-Use Changes on Surface Water Quality—Example from Goriška Brda in Slovenia. Water [Internet]. 2020 Jan 9 [cited 2020 Nov 17];12(1):189. Available from: https://www.mdpi.com/2073-4441/12/1/189

[27] Škerjanec M. Modeliranje obremenitev površinskih in podzemnih voda za oceno scenarijev upravljanja s. 2015;294-302.

[28] Ojsteršek Zorčič P. Izbira in umeščanje ERM ukrepov v vodozbirno območje akumulacijskih jezer. 2015; Available from: http://www.digitalna-knjiznica.bf.uni-lj.si/gozdarstvo/dd_ojstersek_zorcic_polonca.pdf

[29] Glavan M, Flajsig GĆ, Karleuša B, Vučković I. Modelling diffuse and point source pollution risks in the case of transboundary Sotla river basin [Internet]. 2017 [cited 2017 Nov 9]. Available from: http://swat.tamu.edu/media/115777/f1-poster-glavan-et-al-sotla-swat-2017-warsaw.pdf

Chapter 3

Examination of Hydrologic Computer Programs DHM and EDHM

Theodore V. Hromadka II and Prasada Rao

Abstract

The Diffusion Hydrodynamic Model or DHM is a coupled one- and two-dimensional (2-D) surface flow model based upon a diffusion formulation of the well-known Navier–Stokes equations, developed by research hydrologists of the USGS (United States Geological Survey) for use in modeling floodplains and dam-break situations. The Fortran 77 source code and various applications were published in 1987 by the USGS as a Technical Report authored by Hromadka and Yen. The DHM program led to the development of several subsequent computational programs such as the FLO-2D computational model and other similar programs. The original DHM program had a limit of applications to problems with no more than 250 nodes and modeling grids. That limitation was recently removed by a program version named EDHM (Extended DHM), which provides for 9999 nodes and grids. However, the computational code is preserved in order that the baseline code algorithmic procedures are untouched. In this paper, the DHM and EDHM are rigorously compared and examined to identify any variations between the two Fortran codes. It is concluded from this investigation that the two sets of algorithm codes are identical, and outcomes from either program are similar for appropriately sized applications.

Keywords: legacy Fortran codes, computational economy, large scale application, overland flow, flow through a constriction

1. Introduction

Legacy Fortran 77 codes that have been developed in the 1980s continue to have a wide spread audience (for both research and commercial applications) across all the Computational Fluid Dynamics disciplines. Their popularity can be largely attributed to the extensive validation that these models have been subjected to with analytical and experimental data for single and multi-dimensional flows. The trust that the audience have in the end results from these legacy codes, and their ability to meet the user goals are other driving factors for making them popular among the modeling community. Few of the legacy Fortran 77 codes, developed by various groups include MFIX [1] (Open source software for simulating Multiphase Flow with Interphase eXchanges), VOF 2D [2] (Two- dimensional, transient, free-surface incompressible fluid dynamics program), FUN3D [3] (fully unstructured-grid fluid dynamic simulations spanning incompressible flow to transonic flow), LAURA [4] (Langley Aerothermodynamic Upwind Relaxation Algorithm for structured, multi-block, computational aerothermodynamic simulations) and INS3D [5] (incompressible Navier–Stokes equations in three dimensions for steady and transient flows).

Although theoretically, these legacy codes are on a firm footing, computationally, they are uneconomical. When applied over a large scale application, characterized by thousands of nodes, these codes are constrained by varying degrees. The limitations arise partly from (a) the lack of object-oriented tools and the absence of abstract modeling capabilities in Fortran 77 and (b) the required CPU time when these models are applied across large domains. Balancing the accuracy of simulation with acceptable CPU is a crucial element that the current modelers are looking for. Capturing the physics of some flow phenomena necessitates that the equations be applied across the small spatial grid and temporal scales, which can be an issue in applying legacy codes. Since the codes were written (at that time) for the then available computational resources, applying them, as they are, for large scale domains might not be feasible either due to the large CPU time that they need to complete the simulation or because of the array limitations or modifications that need to be made so that the codes can be compiled using the currently available Fortran 77 compilers. Addressing these limitations will result in a 'modernized' version of the legacy codes. Modernization does not mean a better numerical formulation or a more accurate code. It only refers to a computationally efficient code with perhaps a better user interface for input and for visualizing the results through colorful multi-dimensional graphs and tables. In fact, using any modernized code without extensive benchmark testing can lead to erroneous solutions.

The above first limitation was addressed by researchers either by rewriting the entire code from scratch using an object-oriented programming language or by using an incremental approach in which the computationally intensive modules in legacy codes are identified and replaced with their computationally efficient counterparts. For instance, in an implicit finite difference or finite element formulation, where the system of equations are assembled in the form Ax = B, much of the computational time is spent on solving for 'x' [6]. While for the application audience, a solution module merely represents a means to an end of solving the flow equations, for a solver developer, the application is a source of sparse equations to be solved. Using an appropriate solver and a preconditioner can significantly cut down the simulation time, thus partly addressing the limitation.

The advent of high-performance computing tools and their availability to all the audience (based on commodity processors) saw the evolution of legacy Fortran codes to their full or semi-parallel versions, thus addressing the second limitation in legacy codes. There are many advantages to using parallel codes. For an application modeler using a well-written parallel code, the primary advantage is the reduction in the computational time. This reduction is on the order of the number of processors used in the communicator. Parallelization allows a program to be executed by as many processors available within the sub-complex simultaneously, thus facilitating a steep reduction in the computational time required for the solution to converge to the desired simulation time. This reduction enables the code to be applied over domains with millions of nodes or cells, to better capture the physics of the flow. Using standard parallel libraries that use MPI protocols like PETSc [7], ScaLAPACK [8], and BLAS [9], existing serial codes can be converted to parallel codes with reduced effort. Alternatively, highly optimized parallel versions of legacy codes can be written from scratch, which involves higher costs for developing and testing the code.

Diffusion Hydrodynamic Model (DHM) is a legacy model developed in the 1980s for USGS by the first author and his colleague [10]. The report and the Fortran 77 source code are also available at the DHM companion web page http://

diffusionhydrodynamicmodel.com/. The model was extensively tested in the 1980s for various free surface flow scenarios, and its back engine has laid foundation blocks for other popular models like FLO-2D [11]. The DHM was originally developed as the first (or one of the first) 3-D CFD computational programs but was subsequently revised into the form published due to computer limitations of the day. In the last 30 years, DHM has proven to be a practical and reliable tool for predicting two-dimensional surface flow characteristics associated with gradually varied flows and is popular among the hydraulic community. DHM solves a simplified two-dimensional diffusion wave equation, the solution of which is sufficient for many free surface flows that are commonly encountered. For gradually varied flows, the model predicts the values of flow depth and velocity. The model does include any turbulence terms, and it cannot be used for rapidly varying flows. In this work, we address the computational limitations in DHM so that it can be applied over larger domains. The modified DHM, herein, is referred to as Enhanced DHM (EDHM).

The layout of this document is as follows. In Section 2, the flow equations and other salient characteristics in DHM are briefly described. Section 3 lists the computational limitations in DHM. In Section 4, the modifications done in DHM to arrive at EDHM are detailed. Performance tests to validate the reliability of EDHM are discussed in Section 5. Conclusions are presented in Section 6.

2. Overview of DHM

The two-dimensional flow continuity and momentum equations along the X and Y axis (assuming a constant fluid density without sources or sinks in the flow field and hydrostatic pressure distribution) can be written as [10].

$$\frac{\partial H}{\partial t}+\frac{\partial q_x}{\partial x}+\frac{\partial q_y}{\partial y}=0 \tag{1}$$

$$\frac{\partial q_x}{\partial t}+\frac{\partial}{\partial x}\left[\frac{q_x{}^2}{h}\right]+\frac{\partial}{\partial y}\left[\frac{q_x q_y}{h}\right]+gh\left[S_{fx}+\frac{\partial H}{\partial X}\right]=0 \tag{2}$$

$$\frac{\partial q_y}{\partial t}+\frac{\partial}{\partial y}\left[\frac{q_y{}^2}{h}\right]+\frac{\partial}{\partial y}\left[\frac{q_x q_y}{h}\right]+gh\left[S_{fy}+\frac{\partial H}{\partial X}\right]=0 \tag{3}$$

in which q_X, q_y are the unit flow rates along the spatial directions; S_{fx}, S_{fy} represents friction slopes; and h, H, h, g denote flow depth, water surface elevation, and gravity, respectively.

The local and convective acceleration terms can be grouped and Eqs. 2 and 3 are rewritten as

$$m_Z+\left[S_{fz}+\frac{\partial H}{\partial Z}\right]=0, z=x,y \tag{4}$$

where m_Z represents the sum of the first three terms in Eqs. 1,2 divided by gh. Using Manning's formula to calculate the frictional slope, the flow equation can be simplified to

$$q_Z = \frac{1.486}{n} h^{5/3} S_{fz}^{\ 1/2}, \ z = x, y \tag{5}$$

Eq. 5 can be rewritten in the general case as

$$q_Z = -K_Z \frac{\partial H}{\partial Z} - K_Z m_Z, \ \ z = x, y \tag{6}$$

where

$$K_Z = \frac{1.486}{n} h^{5/3} \Bigg/ \left| \frac{\partial H}{\partial S} + m_S \right|^{1/2} \ z = x, y \tag{7}$$

The symbol S in Eq. 7 indicates the flow direction which makes an angle of $\theta = \tan^{-1}\left(q_y / q_x\right)$ with the positive x-direction. By assuming the value of m to be negligible, the diffusion model can be expressed as,

$$q_Z = -K_Z \frac{\partial H}{\partial Z}, \ \ z = x, y \tag{8}$$

Two-dimensional DHM is formulated by substituting Eq. 8 into Eq. 1

$$\frac{\partial H}{\partial t} = \frac{\partial}{\partial X} K_x \frac{\partial H}{\partial X} + \frac{\partial}{\partial y} K_y \frac{\partial H}{\partial y}. \tag{9}$$

If the momentum term groupings were retained, Eq. 9 can be written as

$$\frac{\partial H}{\partial t} = \frac{\partial}{\partial x} K_x \frac{\partial H}{\partial x} + \frac{\partial}{\partial y} K_y \frac{\partial H}{\partial y} + S \tag{10}$$

where

$$S = \frac{\partial}{\partial x}\left(K_x m_x\right) + \frac{\partial}{\partial y}\left(K_y m_y\right),$$

and K_x, K_y are also functions of m_x, m_y respectively.

To maintain continuity in the discussion, while salient aspects of the DHM numerical algorithm are presented here, readers are referred to [10] for a detailed description of the numerical formulation and the input file format. The domain is divided into uniform square grids or cells. For each interior grid, its connectivity with adjacent grids along the North, East, South, and West directions is specified. For grids that are on the boundaries, '0' is specified along the directions that do not have adjacent nodes. The flow equation is solved using the nodal domain integration method. Apart from the grid connectors, at the center of each cell, the required input variables are the roughness value, ground elevation, initial flow depth. The number of inflow cells and the inflow hydrograph at each of them need to be inputted. The number and the outflow boundary cell numbers should be identified. Since the formulation is explicit, the choice of time step (Δt) is limited by the Courant-Friedrich-Lewy stability condition. Starting from time = 0, the explicit solution is marched in the direction of time, until the required transient time level is reached. DHM gives the option of printing the output variables at any time level. The output at the center of each cell includes the flow depth, elevation, and flow velocities along the four directions.

3. Computational limitations in DHM

For modeling flows over large sizes domain, the two primary shortcomings in DHM are

a. The maximum number of cells (nodes) that can be accommodated in DHM is limited to 250

b. Inflow and outflow boundary nodes are limited to 10

Both these limitations were largely due to the computational resources that were available to the developers in the 1980's. Application of DHM over large flow domains would require using a higher number of nodes in the computational domain, warranting modifications to DHM, as discussed next.

4. Features in extended DHM (EDHM)

The changes made in DHM can be grouped into three categories, as detailed below. No changes were made to the format or structure of variables input and output files.

4.1 Major enhancements

a. Increased the array size of variables (FP,FC) that stores the location of cells and the initial depth, average elevation, roughness coefficient values, velocity (VEL), maximum water depth, and the corresponding time at the cell (DMAX, TIMEX) from 250 to 9999.

b. Increased the array size of variables that store the stage curve data for the channel (NOSTA, STA) from 10 to 99.

c. Increased effective rainfall intensity data pairs from 10 to 99.

d. Increased the array size of variables relating to the inflow and outflow hydrograph nodes (KIN, KOUT), depth of the specified stage-discharge curve (HOUT), inflow boundary condition nodal points (KINP) and the inflow hydrograph details (HP) from 10 to 99

e. Increased of the array which stores the nodal points where outflow hydrographs are being printed (NODFX, NODCC) from 50 to 99

The changes in the above array sizes were done in the main code and the associated subroutines FLOODC, QFP, QFC, and CHANPL.

DHM		EDHM	
Line #	Content	Line #	Content
12	COMMON/BLK 1/FP(250,8),FC(250,6)	12	COMMON/BLK 1/FP(9999,8),FC(9999,6)
14	COMMON/BLK 2/KIN(10),H(10,15,2), KOUT(10),HOUT(10,15,3)	14	COMMON/BLK 2/KIN(99),H(99,15,2), KOUT(99),HOUT(99,15,3)
16	COMMON/BLK 3/NOSTA(10),STA(10,15,2),NODFX(50)	16	COMMON/BLK 3/NOSTA(99),STA(99,15,2),NODFX(99)
18	COMMON/BLK 4/DMAX(250,2),TIMEX(250,2)	18	COMMON/BLK 4/DMAX(9999,2),TIMEX(9999,2)
20	COMMON/BLK 5/KINP(10),HP(10,15,2)	20	COMMON/BLK 5/KINP(99),HP(99,15,2)
26	DIMENSION NODDC(50),VEL(250,4),R(10,2),Q(4)	26	DIMENSION NODDC(99),VEL(9999,4),R(99,2),Q(4)
271	FORMAT(10X,5I4,1X,F6.4,2X,F6.1,1X,F5.1)	271	FORMAT(10X,5I5,1X,F6.4,2X,F6.1,1X,F5.1)
281	FORMAT(/,10X,'INFLOW HYDROGRAPH AT NODE #',I3,/,	281	FORMAT(/,10X,'INFLOW HYDROGRAPH AT NODE #',I4,/,
291	FORMAT(10X,I3,1X,I3)	291	FORMAT(10X,I4,1X,I4)
297	FORMAT(10X,I3 ,2X,F5.4,1X,F7.1,1X,F7.1,1X,F7.1,5X,F7.1)	297	FORMAT(10X,I4 ,2X,F5.4,1X,F7.1,1X,F7.1,1X,F7.1,5X,F7.1)
303	FORMAT(10X,'OUTFLOW NODE # ',I3,	303	FORMAT(10X,'OUTFLOW NODE # ',I4,
311	FORMAT(/,10X,'STAGE CURVE AT NODE #',I3,/,	311	FORMAT(/,10X,'STAGE CURVE AT NODE #',I4,/,
333	FORMAT(10X,'INFLOW RATE AT NODE ',I3,' IS EQUAL TO ',F10.2)	333	FORMAT(10X,'INFLOW RATE AT NODE ',I4,' IS EQUAL TO ',F10.2)
335	FORMAT(/,5X,'NODE',7X,10(I3,8X))	335	FORMAT(/,5X,'NODE',7X,10(I4,7X))
357	FORMAT(10X,'OUTFLOW RATE AT NODE ',I3,' IS EQUAL TO ',F10.2)	357	FORMAT(10X,'OUTFLOW RATE AT NODE ',I4,' IS EQUAL TO ',F10.2)
1275	COMMON/BLK 1/FP(250,8),FC(250,6)	1275	COMMON/BLK 1/FP(9999,8),FC(9999,6)
1277	COMMON/BLK 2/KIN(10),H(10,15,2), KOUT(10),HOUT(10,15,3)	1277	COMMON/BLK 2/KIN(99),H(99,15,2), KOUT(99),HOUT(99,15,3)
1279	COMMON/BLK 3/NOSTA(10),STA(10,15,2),NODFX(50)	1279	COMMON/BLK 3/NOSTA(99),STA(99,15,2),NODFX(99)
1281	COMMON/BLK 4/DMAX(250,2),TIMEX(250,2)	1281	COMMON/BLK 4/DMAX(9999,2),TIMEX(9999,2)
1567	COMMON/BLK 1/FP(250,8),FC(250,6)	1567	COMMON/BLK 1/FP(9999,8),FC(9999,6)
1647	COMMON/BLK 1/FP(250,8),FC(250,6)	1647	COMMON/BLK 1/FP(9999,8),FC(9999,6)

DOI: http://dx.doi.org/10.5772/intechopen.94283

DHM		EDHM	
Line #	Content	Line #	Content
1727	COMMON/BLK 1/FP(250,8),FC(250,6)	1727	COMMON/BLK 1/FP(9999,8),FC(9999,6)

Table 1.
Detailed listing of the differences between the DHM and EDHM source codes.

4.2 Minor enhancements

The two minor enhancements that were made in DHM code are (a) to accommodate the increased number of cells in EDHM, the fixed format output descriptor has been expanded by one digit and (b) to better align the variables in the output file, the inter variable spacing was decreased by one digit. A detailed listing of all the major and minor changes made in the DHM source code, along with the corresponding line numbers, is shown in **Table 1**.

4.3 Compiler details

After reviewing the currently available compilers in Windows for Fortran 77 codes, we have chosen the Intel Fortran Compiler within the Microsoft Visual Studio integrated development environment (IDE) to make the enhancements in DHM and for generating the.EXE file. This interface is ideal to debug and execute Fortran 77 programs. The compiler can optimize the performance of source codes for Intel CPUs. It offers broad support for current and previous Fortran standards and also tools by which a robust, high-performance code can be created in serial and parallel environments. The Math Kernel Library (Intel MKL) and the Debugger tools in the compiler, creates a solid foundation for building robust, high-performance codes. The end executable file (.EXE) although optimized for Intel CPUs, can also run on x86 compatible CPUs such as those from AMD.

5. Application of EDHM

EDHM is qualitatively and quantitatively evaluated by comparing its results with the output from DHM, for two test simulations. The focus was to check if the EDHM solution resembles DHM output for these two cases for varying inflow and other model parameters.

Case 1: Flow in a transition.

Open channel flow through a linear contraction under the framework of two-dimensional flow is a common phenomenon and has drawn the attention of many experimental and numerical studies. The flow characteristics in the contraction depend on the Froude number at the upstream end. Flow in a contraction has acted as a benchmark simulation in investigations [12] that compared the performance of various CFD and hydraulic models. **Figure 1** is the definition sketch of the test problem. The rectangular channel is 380 ft. long and 260 ft. wide. The constriction

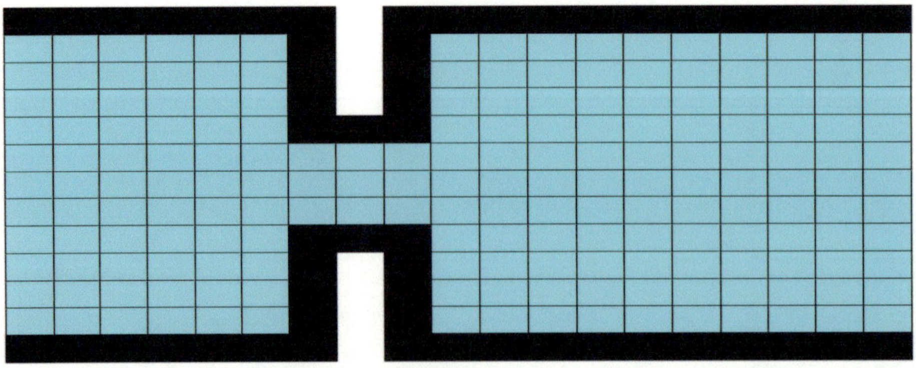

Figure 1.
Definition sketch of the test problem (channel length = 380 ft., channel width = 260 ft., cell size = 20 ft).

portion of the channel is 60 ft. x 60 ft. The channel length before and after constriction is 120 ft. and 200 ft., respectively. The cell size in the domain is 20 ft., and the total number of cells are 239. **Figure 2** illustrates the cell numbers in the domain. The elevation of cells along the north and south boundaries was assigned a high value to physically denote that they are walls. The flow is confined within these boundaries. At the upstream end, cells 3-11 (nine cells) were specified with a constant inflow of 33.33 cfs. At the downstream end, a free outfall boundary was specified. The transient simulation was carried out until time = 1 hour.

Figure 3 plots the water depth profile for the two models along the channel centerline at time = 0.9 hours for a channel bottom Manning's roughness value of 0.015. The close agreement of results gives confidence that the changes made to arrive at EDHM did not lead to different output data. The analysis over varying inflow discharge and bottom roughness values did not change the trend of the results. **Figures 4** and **5** compare the depth profile for the roughness coefficient of 0.024 and 0.04. A similar agreement in output data (including velocity components) was observed across other longitudinal sections.

Case 2: Overland flow.

Overland flow over a hill slope generated by a rainfall event is characterized by varying hydraulic properties, roughness values, topography, and physical features in the domain. For predicting the hydraulic and hydrologic properties of flow, various models that solve a range of equations from a one-dimensional hydrodynamic equation for homogeneous place surfaces [13, 14] to 2D full non-linear shallow water Equations [15] have been applied. **Figure 6** shows the flow domain. The number of cells in the domain is 56 and are 30 ft. in size. The roughness value ranged between 0.015 to 0.03Inflow hydrograph (**Figure** 7) was applied at cells 1 and 15. The transient model was run until time = 2 hours. The elevation drop along the north and south end of the domain is 26 ft., and the drop along the west and east boundaries is 5 ft. Cells 42 to 56 were specified as critical depth outflow

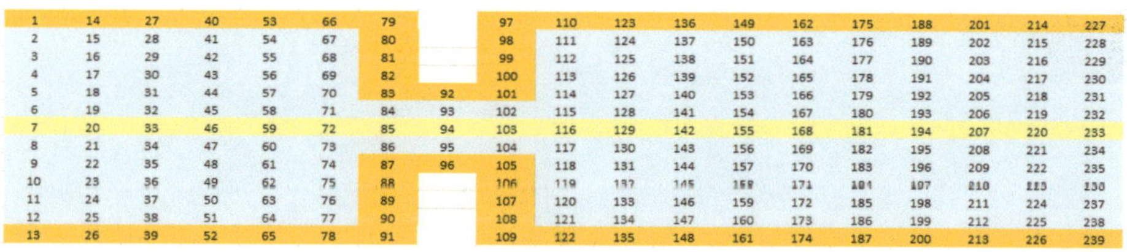

Figure 2.
Flow domain with the cell numbers. The wall boundaries are identified by orange-colored cells. There are 239 cells in the domain. The centerline cells (7,20,..220,233) where the depth profiles are compared are highlighted.

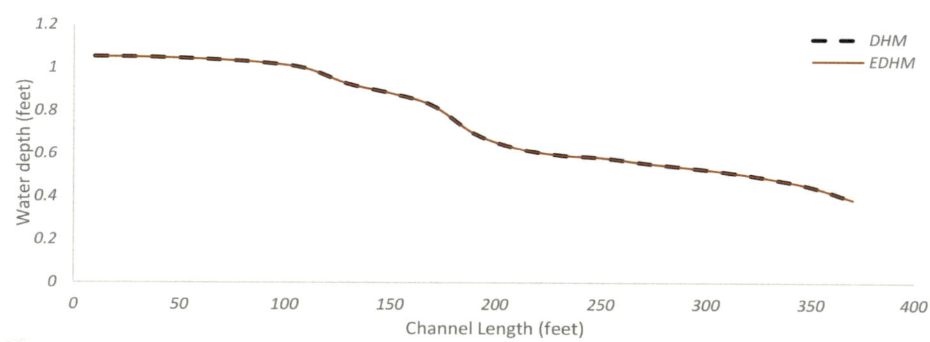

Figure 3.
Comparison of depth profiles along the centerline of the channel at time = 0.9 hours (roughness = 0.015).

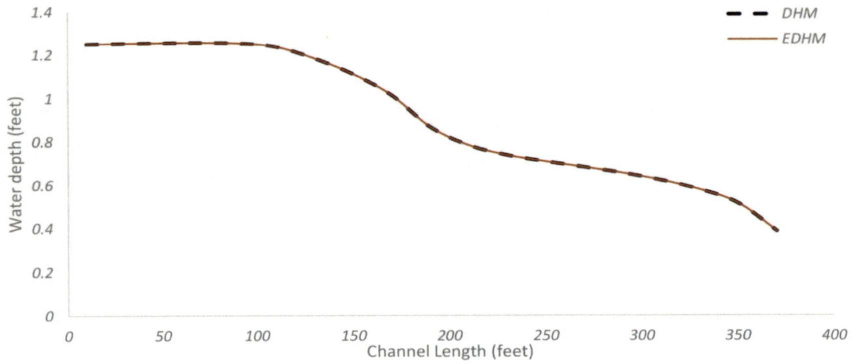

Figure 4.
Comparison of depth profiles along the centerline of the channel at time = 0.9 hours (roughness = 0.024).

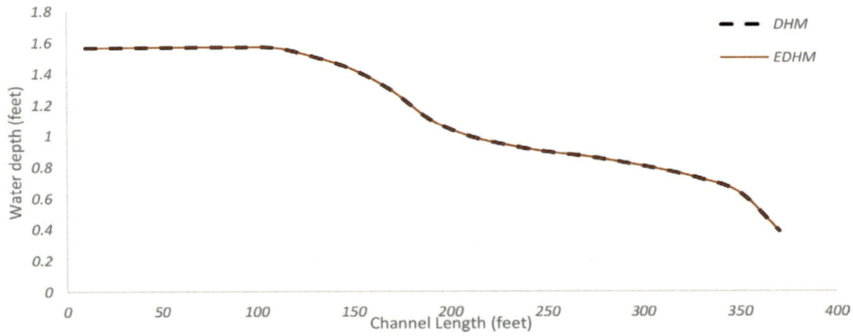

Figure 5.
Comparison of depth profiles along the centerline of the channel at time = 0.9 hours (roughness = 0.04).

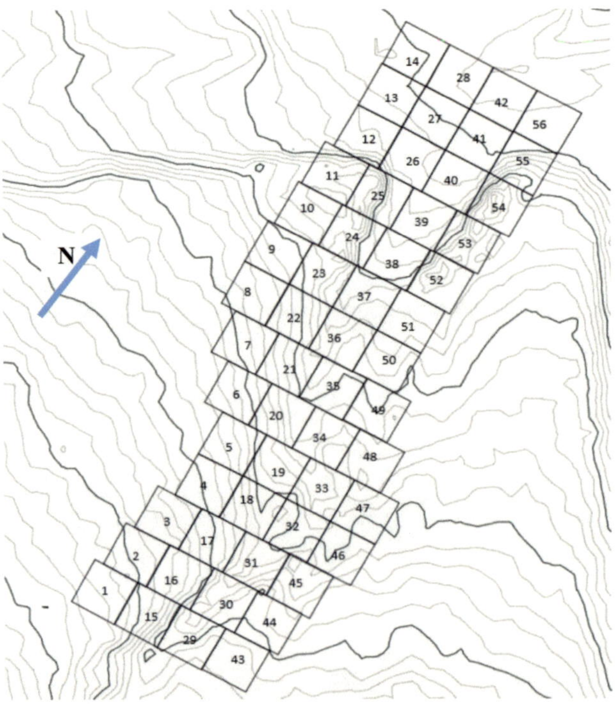

Figure 6.
Overland flow domain with the cell numbers. The domain has 56 cells.

nodes. **Figures 8** and **9** compare the transient depth profiles from DHM and EDHM (time = 0 to 2 hours) at cells 36 and 26, respectively. The trend of the depth and velocity values at other cells also indicated the close agreement of flow data from the two models.

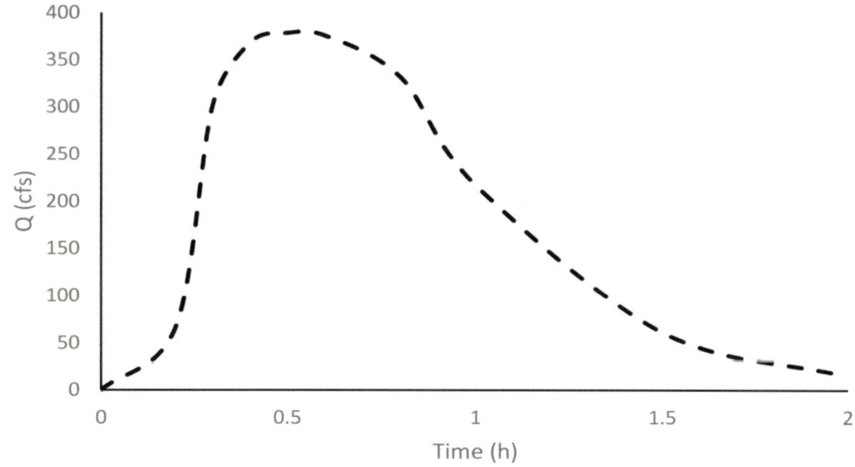

Figure 7.
Inflow hydrograph data at cells 1 and 15.

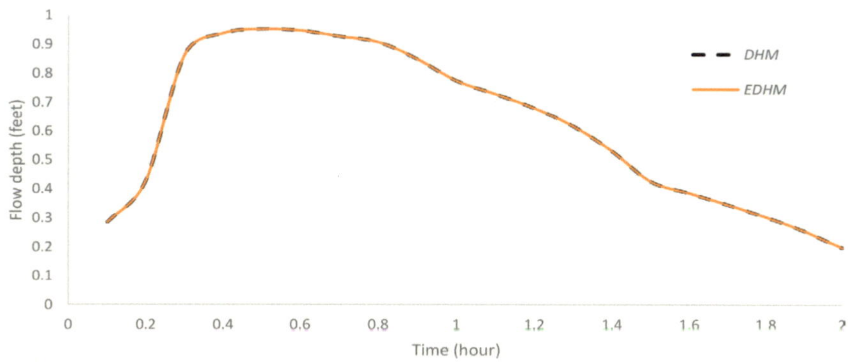

Figure 8.
Transient depth profiles at cell 36.

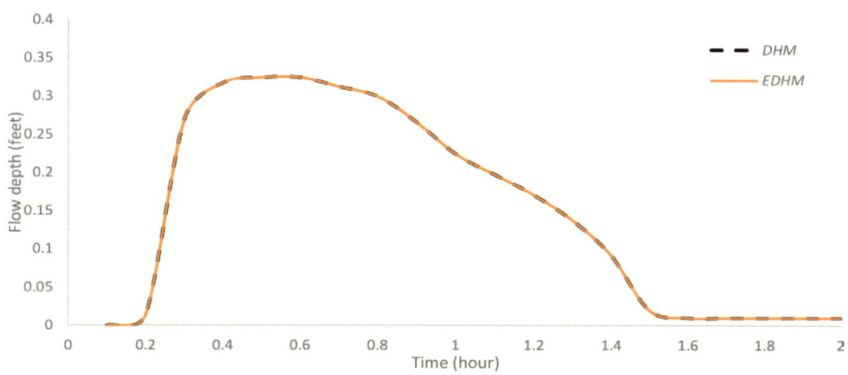

Figure 9.
Transient depth profiles at cell 26.

6. Conclusions

USGS Diffusion Hydrodynamic Model is a legacy Fortran 77 that has been widely applied for multiple one and two-dimensional flow scenarios. The computational limitations in the model which prevent its application over large domains have been addressed in this paper. The enhanced model can accommodate 9999 cells and 99 inflow and outflow nodes. Based on the analysis that was carried out and the close agreement of the results between the two models gives confidence in the reliability of the extended model. Current findings encourage future development of parallel EDHM in order to reduce the computational time.

Author details

Theodore V. Hromadka II[1] and Prasada Rao[2*]

1 Department of Engineering-Mathematics, United States Military Academy, West Point, NY, United States of America

2 Department of Civil and Environmental Engineering, California State University, Fullerton, CA, United States of America

*Address all correspondence to: mprasadarao@fullerton.edu

References

[1] Syamlal, M. MFIX documentation. User's manual, U.S. Dept. of Energy, DOE/METC-95/1013, DE95000031, https://mfix.netl.doe.gov/doc/mfix/20.1.0/, 2004.

[2] Torrey MD, Cloutman LD, Mjolsness RC and Hirt, CW.NASA-VOF2D: A Computer Program for Incompressible Flows with Free Surfaces.LA-10612-MS(December 1985) http://www.oecd-nea.org/tools/abstract/detail/nesc9644/

[3] Nielsen E, Park MA, Rumsey CL, Thomas JL, and Wood WA. FUN3D Manual: 13.0," NASA TM-2016-219330, Langley Research Center, Aug. 2016. https://fun3d.larc.nasa.gov/

[4] Mazaheri A, Gnoffo PA, Johnston CO, and Kleb B. LAURA Users Manual: 5.5-65135, NASA TM-2013-217800, February 2013. https://ntrs.nasa.gov/search.jsp?R=20130009520

[5] Rogers, SE, Kwak D, and Chang JL. INS3D-an Incompressible Navier-Stokes Code in Generalized Three-Dimensional Coordinates, NASA TM-100012. 1987. https://ntrs.nasa.gov/archive/nasa/casi.ntrs.nasa.gov/19880003609.pdf

[6] Eriksson K, Estep D, Hansbo P. Johnson C. Computational Differential Equations: Cambridge Univ. Press; 1996

[7] Balay S, Brown J, Buschelman K, Eijkhout V, Gropp WD, Kaushik D, Knepley MG, McInnes LC, Smith BF, Zhang H. PETSc users manual. Tech. Rep. ANL-95/11 - Revision 3.4, Argonne National Laboratory, 2013.

[8] Blackford LS, Choi J, Cleary A, D'Azevedo E, Demmel J, Dhillon I, et al. Guide. SIAM; 1997

[9] Chtchelkanova A, Gunnels J, Morrow G, Overfelt J, van de Geijn RA. Parallel implementation of BLAS: General techniques for level 3 BLAS. In: TR-95-40, Department of Computer Sciences, University of Texas. Oct. 1995

[10] Hromadka II TV, Yen CC. A diffusion hydrodynamic model. Water resources investigations report. U.S. geological survey; 1987: 87– 4137. https://pubs.er.usgs.gov/publication/wri874137

[11] O'Brien JS, Julien PY, Fullerton WT. Two-dimensional water flood and mudflow simulation. Journal of Hydraulic Engineering ASCE. 1993;**119**:244-261

[12] Paudel M, Roman SB, Pritchard J. A Comparative Study of HEC-RAS 2D, TUFLOW, & Mike 21, Presented at ASFPM 2016 Annual National Conference. MI: Grand Rapids; 2016

[13] Liggett JA, Woolhiser DA. Difference solutions of the shallow-water equation, J. Eng. Mech. Div. Am. Soc. Civ. Eng. 1967;**9J**(EM2):39-71

[14] Akan AO, Yen BC. Mathematical modeling of shallow water flow over porous media. J. Hydraul. Div. Am. Soc. Civ. Eng. 1981;**W7**(HY4):479-494

[15] Singh J, Altinakar MS, Ding Y. Numerical modeling of rainfall-generated overland flow using non-linear shallow-water equations. Journal of Hydrologic Engineering. 2015;**20**(8). DOI: 04014089, 10.1061/(ASCE)HE.1943-5584.0001124

Chapter 4

Ecohydrology: An Integrative Sustainability Science

Maciej Zalewski

Abstract

The dynamic of the water cycle in catchments is determined by climate, geology, geomorphology, plant cover ad modified by agriculture, urbanisation, industrial development and hydroengineering infrastructure. Up until the end of the 20th century, water management was dominated by a mechanistic approach, focused on the elimination of threats such as floods and droughts and providing resources for the society with little to no regard for the impact this approach had on the ecosystem. Highlighting of water as a key driver of ecosystem dynamics, and further ecohydrology which highlights water/biota interactions from molecular to catchment scale provide a new perspective, new tools and new systemic solutions for enhancement of catchment sustainability potential WBSRCE (consisting of 5 elements: Water, Biodiversity, Ecosystem Services for Society, Resilience and Culture and Education).

Keywords: ecohydrology, sustainability potential, engineering harmony, water, management

1. Introduction: why ecohydrology becoming one of the key for sustainable biosphere, water and food

"We are living in the Anthropocene Era when almost 80% of our usable ecosphere, has been conditioned, converted, and consumed by humans, usually without understanding the full consequences of our actions" [1].

There is an increasing the number of the scientific evidences from molecular, ecosystem up to global scale, that the exponential growth of human population and acceleration of consumption in Anthropocene resulted in the declining the ecological and regenerative potential of the planet Earth, expressed by the "ecological footprint", which recently is above 1.7 [2]. This accelerated changes of biosphere can be described in two dimensions: first cumulative - synergic amplification many impacts (deforestation +pollution+ river channelization ect.) The second one - long term slow changes e.g. catchment urbanisation, industrialisation, transport development, emission of pollutants. Both create dramatic consequences: reduces water retentiveness and increase stochastic character of water cycle – floods, droughts, landslides and interconnected degradation of biogeochemical cycles – carbon, nitrogen, phosphorus and as further consequence the loss of soil fertility. All above processes increase abiotic disturbances strength for biota and decline biological productivity and biodiversity in catchment scale, which in turn negatively effects water quantity and quality. If we continue such "business as usually" deadly spiralling of

IntechOpen

Figure 1.
Example of a drastic reduction in complexity of agricultural landscape – tree rows and land/water ecotone buffer zones, resulting in in soil drying, loss of organic matter due to aeolian erosion and surface flow.

the human impact, in which water is key driver, this can lead to decline and even extinction of civilisation due to mass migrations local, regional and global conflicts.

The one of the major reasons of such decline local regional and global sustainability potential, is that the increasing strength and complexity of the of pressures and interactions between Man and the environment are still not sufficiently reduced and compensated by scientific, technological, progress as far as its translation best current knowledge into society environmental consciousness legal frameworks and policy.

The example of the negative consequences of the recent sociocentric/mechanistic paradigm and lack of understanding the complexity ecohydrological process in recent water management and agricultural policy, has been introduced on **Figure 1**. The Water Framework Directive of European Commission require the achievement the good ecological status by EU member countries however, the agricultural policy by providing the financial support proportional to area of cultivated land has been indirectly encouraging the farmers to maximise such area even by elimination of land/water ecotones and tree rows - shelter belts, which are important buffers reducing nutrients fluxes of agricultural origin [3, 4]. Additionally they are reducing of wind speed, improve soil moisture and agricultural yield [5]. Such simplified agricultural land dramatically increase loads of phosphorus and nitrogen from non-point source pollutions, which in case of the Baltic Sea create more than 40% of the load. Recent agricultural policy focused only on maximisation of yield also accelerate organic matter loss and necessity to compensate reduced soil fertility by more intensive artificial fertilisers use, both of which further intensify eutrophication of lakes, reservoirs and costal zones.

2. Technogarden vs. sustainable anthropobiosphere

When the regenerative potential of the Earth is much below its equilibrium stage, the fundamental question for Humanity is how to achieve desirable safe prosperous future and whether our goal should be a Sustainable Biosphere or we should reconstruct Earth System as Technogarden.

The key assumption of technogarden scenario is that the recent and further technological progress based on experience gained at spaceship expeditions will be sufficiently fast and efficient to create on Earth a technogarden where humanity or a fraction of humanity will persist. Unfortunately the unsuccessful experiments in creation of sustainable technogardens [6] indicate that to maintain homeostatic equilibrium even at the ecosystems level, which means provision of basic ecosystem services: water, food, health and social interactions there is nothing better than the natural or seminatural ecosystems within the catchment, which maintains, self-regulatory water/nutrients cycling thus bioproductivity and biodiversity, thus the positive future we want.

3. The urgent need of paradigm change from sociocentric/mechanistic to evolutionary/ecosystemic

Up to now the science and technology operating in framework of Sociocentric/Mechanistic paradigm has been used as a tools for the intensification of exploitation of natural resources. However recently to achieve sustainable future there is an urgent need to change such paradigm into Evolutionary/Ecosystemic. This new paradigm has to be based on a profound understanding of how evolution determines fundamental ecological processes, first and foremost cycling of water and nutrients and next the whole range of water-biota interplay in different ecosystems, which is a basic role of Ecohydrology. Secondly the above understanding has to be used to develop deductive and inductive models of processes which are based on integration of knowledge form various disciplines of environmental sciences. Testing of the above empirical models provide an opportunities for discovery of new, emerging properties of the systems, which in turn can be translated into innovative methods and systemic solutions [7]. The above idea becomes a background of the evolution/ecosystemic paradigm [8], where Man is a component of Nature and has to obey the roles determined by biological evolution to achieve suitability. This also means that to achieve sustainable future we have to stimulate to a greater extent the integrative, transdisciplinary, environmental science which has the potential not only to highlight the complexity and specifics of the Man –Environment interactions at different continents and various cultural contexts, but also by empirically testing the highly advanced models of ecohydrological processes, generate innovative methods and systemic solutions for enhancement of sustainability potential WBSRCE of catchments [7]. WBRSCE expresses the urgent need for proactive management in face of the crisis generated by degradation and overexploitation of Biosphere. The key assumption is that in the recent stage of Anthropocene, it is not enough to conserve the nature, but much more intensive action is necessary towards reversing degradation of biocenosis, not only by reduction of impact e.g. by circular economy but, also by the regulation of fundamental ecological process such as water nutrients cycling expressed by a plethora of feedbacks between water and biota. The profound understanding of those relations become background for parallel enhancement of key parameters which determine catchment sustainability –Water, Biodiversity, Services from ecosystems for society, Resilience to climate and various anthropogenic impacts and Culture and education - WBSRCE.

4. Why water and its interplay with ecosystems? genesis of ecohydrology- dual and mutual nature

Ecohydrology emerged in the last decades of the XX Century as a result of the parallel efforts of hydrologists, hydrobiologists, botanists and plant physiologists. The roots of the aquatic phase of ecohydrology which integrates in the framework of physics laws, hydrology and ecology has to be considered in the Abiotic-Biotic

Regulatory Continuum (ABRC, **Figure 2**) [9, 10], which introduce changes of hierarchy of abiotic (hydrology, thermodynamics) vs. biotic drivers (competition, predation, adaptive life strategies) along the river continuum at different climatic regions. One of the inspirations to develop this model was the debate between ecologists on density dependent and density independent regulation factors of populations [11]. ABRC model indicates a gradient of density dependent and independent regulation, which in rivers depends on physics – Bernoullie's principle and temperature determinant of oxygen availability, energy flow, nutrients cycling and biological productivity.

The model also provides a predictive tool e.g. in the abioticaly regulated rivers of boreal zone, to enhance fish populations and stocking efficiency it is necessary to reduce hydraulic stress and energy expenditure in trophy limited ecosystems. On the other hand in biotical regulated tropical rivers there is a necessity to increase trophy potential to reduce competition for food and improve spatial diversity of habitat to reduce predator pressures. This model provided a framework for the FAO UN EIFAC programme Habitat Modification and Freshwater Fisheries. In the next important stage, the ABRC model supported the development of Ecohydrology UNESCO Intergovernmental Hydrological Programme. The expression of Ecohydrology underline that abiotic factors (hydrology, temperature) are primary drivers of ecosystem functioning as far as biota is filling the template created by abiotic factors [12]. An important step was the UNESCO MAB Programme "Role of the land/water ecotones in landscape management and restoration [3, 13] and further regulation of hydrological dynamics of reservoir for shaping the trophic cascade towards mitigation of eutrophication symptoms [14], the environmental flow in the face of global climate changes [15] and the Ecological Engineering idea [16, 17] were supportive to integrate the above puzzles in the international expert team involved in the framework of UNESCO International Hydrological Programme [18, 19] and especially for development of the key idea of **Ecohydrology** as a transdisciplinary sustainability science - **using ecosystem properties and processes as a tool for**

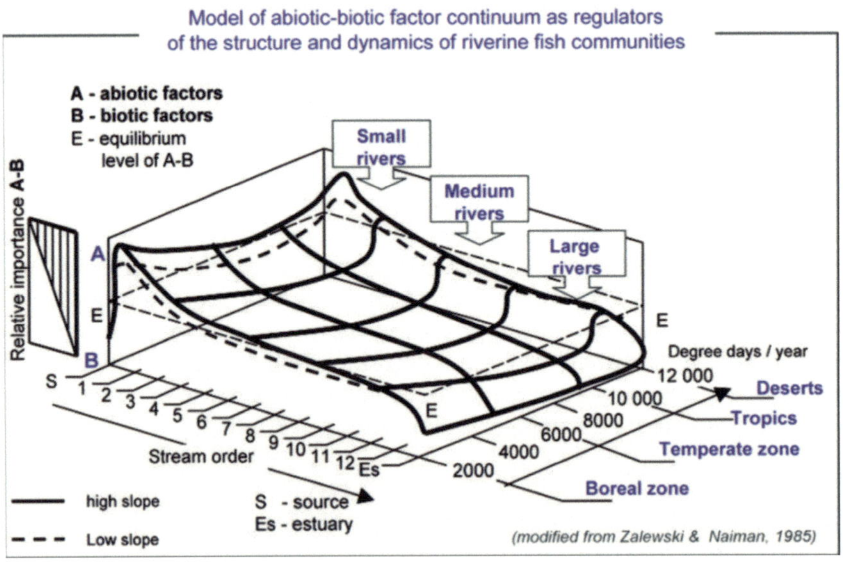

Figure 2.
Deductive model of changes in hierarchy of abiotic and biotic drivers along the river continuum and the temperature gradient: The stream order determines the hydrodynamics and energy demand for trophic processes (hydrology), whereas the temperature determines the metabolic rate, growth rate and bioproductivity (ecology). Water quality, bioproductivity and biodiversity depend on the nutrient spiralling rate, flood pulses and the relation of energy intake from food and the energy expense determined by the above processes.

holistic catchment management [19] and enhancement of sustainability potential WBSRC [20]. Independently the terrestrial phase of ecohydrology has been deepening our understanding of water/plants/soil interactions [21–24] - which is crucial for catchment management especially for shaping terrestrial ecosystems distribution and structures for concordant enhancement of resilience to climate change, reduction non-point source pollution, ecosystem services and biodiversity. The synthetic model which introduced dependence of terrestrial ecosystems bioproductivity and biodiversity and in consequence its sustainability status on the catchment water retentiveness and temperature was introduced in a model build on empirical data [25]: Water, Temperature, Bioproductivity and Biodiversity (WTBBS) (**Figure 3**). This model indicate that reduction of catchment retentiveness by reduction of forest cover, ecotone shelter belts between fields, elimination land-water ecotones and streams/rivers channelization (see **Figure 1**) not only reduce organic matter in soils due to water and wind erosion but generate eutrophication of inland and coastal waters and in situation of climate changes accelerating nutrient cycling indirectly reduce biodiversity and bioproductivity.

A further crucial step for implementation of Ecohydrology was developed on modelling large scale processes especially in costal zones and sea [26–29], analysis of the water biota interplay at different ecosystems [30, 31], employment of molecular

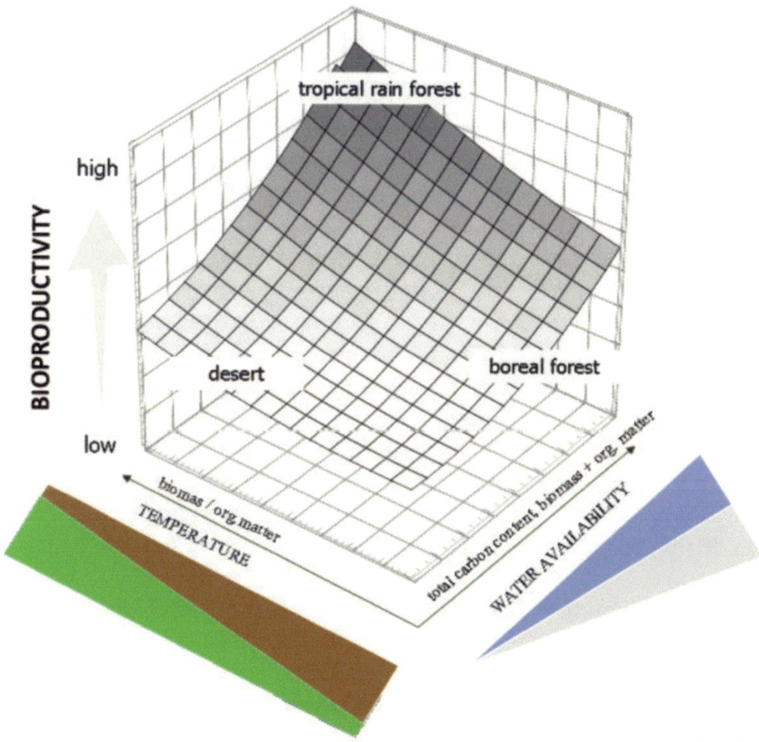

Figure 3.
A deductive model based on empirical data gathered worldwide. The amount of accessible water determines the ecosystem capability to accumulate organic carbon, whereas temperature determines carbon allocation between organic matter (in soil) and living tissues (plants). In low temperature zones most of the carbon is allocated in organic matter, because of low temperatures block decomposition processes and nutrient cycling. On the contrary, high temperatures, e.g. a tropical forest, most of the carbon is allocated in living tissues or organisms, because in such conditions favour high decomposition rate (bioproductivity). The model shows that in conditions of high water accessibility and high temperature, the recirculation rate allows a high bioproductivity and biodiversity. Worldwide long-term consequences of acceleration of river outflow which is occurring in the catchment because of deforestation and river canalization reduces accumulation of carbon, bioproductivity and biodiversity. Ecohydrology systemic solutions provide framework how to reverse these processes.

biology methods for freshwater ecosystem diagnosis [32] and next by translation of the integrative understanding of water biota interplay into ecosystemic biotechnologies [33–35] testing and development Ecohydrological Nature Based Solutions for Water in catchment scale [36]. Also the understanding of society priorities [37], becomes crucial for its involvement in reduction of dispersed impacts and broad range of activities improving WBSRCE. All above efforts especially developed in framework of UNESCO Man and Biosphere and International Hydrological Programme generate inspiration and provides certain prototype for development of the concept of Nature Based Solutions for Water [38].

The parallel step which has been deepening the understanding the various hierarchy of drivers and specific properties of the ecosystems vs. societies priorities was generated by the testing the best current Ecohydrological wisdom and biotechnologies into African conditions [39–41] and use indigenous knowledge in Ecohydrological solutions [42] and analysis of processes and versus human impacts in catchment perspective [43].

5. Principles of ecohydrology as framework for action

Water is the common denominator and abiotic factors: hydrology and temperature are of primary importance in shaping the biological structure and processes within ecosystems for all types of climatic biogeochemical and ecological processes, thus water mezocycle within the catchment provides the best operational template for regulation of water-biota interplay towards enhancement of ecological potential and to achieve desirable status of the ecosystem and sustainable use of its resources. That is why the framework for elaboration and implementation of systemic approach for innovative Ecohydrological Nature Based Solutions (EH NBS) are three principles of Ecohydrology [25]:

I. **Hydrological principle** of Ecohydrology focus on: quantification of hydrological cycle with the special emphasis on the range and dynamics and its modifications due to human impacts, considering the geomorphological structure of the catchment, soil quality (flood, ground water recharge and drought vulnerability), erosion, identification and distribution of various forms of impact e.g. point sources pollution vs. non-point source pollution, urbanisation, transportation pathways. Moreover the timing of river pulses vs. water resources demand e.g. for agriculture and vulnerability to pollution during low flows periods (**Figure 4.**)

II. **Ecological principle** analysis the distribution and ecological potential (WBSR) of pristine – to be protected and novel [44] ecosystems, where novel are a result of the secondary succession and can be a subject of structure changes for processes regulation towards enhancement of carrying capacity (WBSR). **Figure 4.II.** introduce the forests distribution in upper Pilica River catchment, which is important for increasing water retentiveness and groundwater recharge. The enhancement of floodplain phosphorus absorbing capacity and biomass yield begins from analysis of floodplain plants community which combined with groundwater level provides opportunity to replace flooded meadow in 40% by bioenergy willow plantation doubling phosphorus absorbing capacity and profitability of agricultural yield of energy willow from floodplain [45].

III. **Ecological Engineering principle** focused on three types of solutions:

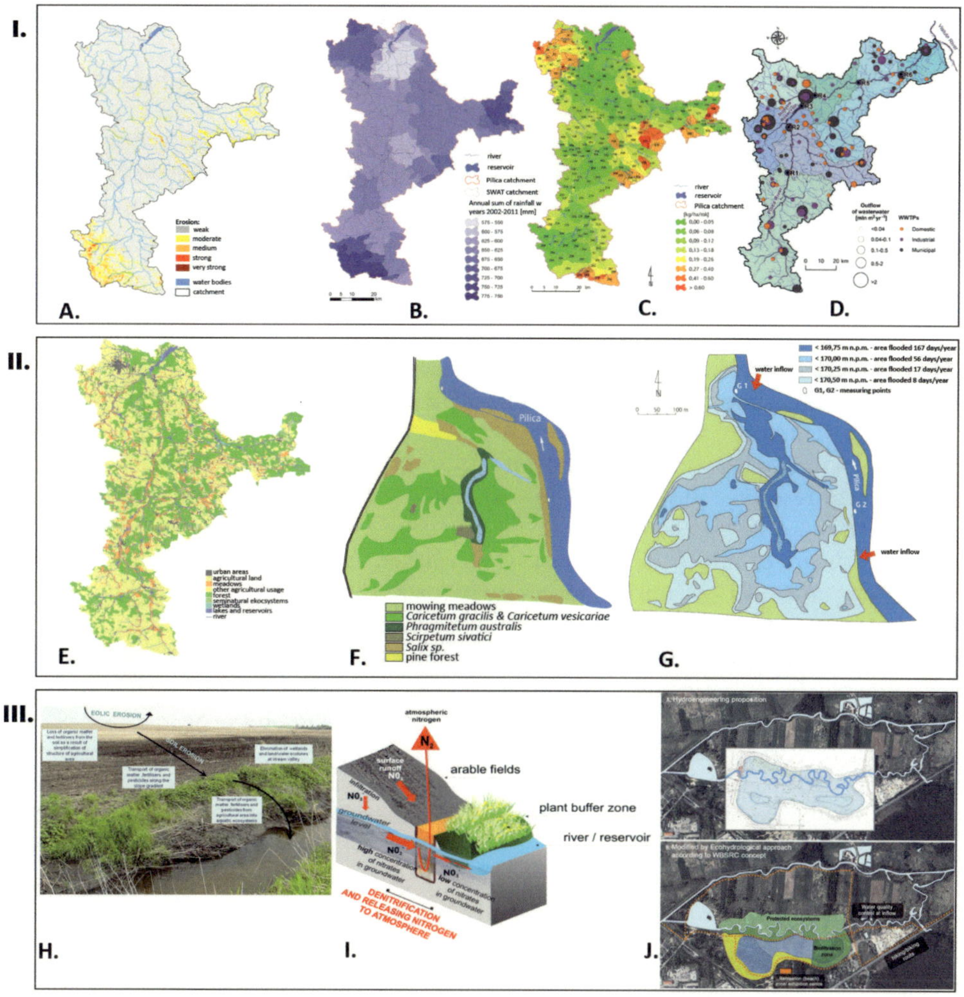

Figure 4.
Principles of Ecohydrology. I principle, hydrological: A. map of water erosion special distribution, describing geomorphological conditions. I.B. distribution of rainfall in the catchment. I.C. SWAT model of special distribution of phosphorus load from non-point source pollution. I.D. distribution of small treatment plants, contributing to total nutrient load in the catchment. II principle, ecological: E. distribution of various land use in the catchment. II. F. Distribution of plant cover on the floodplain corresponding with the time of flooding, allowing for bioenergy plantations of willows in the floodplain, allowing the absorption of phosphorus into plant tissues and increasing the self-purification efficiency by 40% (400 kg of P absorbed). II.G. water level and the time the floodplain stays underwater. III principle, ecological engineering: III. H. Example of a drastic reduction in complexity of agricultural landscape – Tree rows and land/water ecotone buffer zones, resulting in in soil drying, loss of organic matter due to aeolian erosion and surface flow. III. I. the high efficiency ecotone zone with a denitrification barrier for decreasing the nitrogen load from agricultural landscape into ground and surface water. III.J. Ecohydrological alternative solution to proposed by Hydroengieeners reservoir design which blocks the river continuum by eliminating a section of natural, meandering river of high biodiversity. Moreover observed periods of high concentration of phosphorus loads can stimulate toxic algal blooms and eliminate the recreational use of the reservoir. Proposed Ecohydrological approach maintaining the river continuum of the pristine, meandering river and enhancing the catchment sustainability potential by improvement of: W - water quality by eliminating toxic algal blooms, biodiversity, B – Biodiversity by increasing habitat diversity, S – Increasing recreation, R – Increasing river system resilience to climate change by increasing the river valley retentiveness, CE – Citizen science, sustainability consciousness and participation of society.

1. Development and implementation of innovative EHNBS tools with special emphasis on the"dual regulation" – regulation of water cycle/biota interaction from molecular to landscape scale, by shaping biota and regulating biotic processes and vice versa, enhancing biota by regulating hydrology [20, 46].

2. Harmonisation hydroengineering with EHNBS in to hybrid systems [35, 47].

3. Integration of EHNBS hydroengineering and hybrid solutions for synergy at catchment scale [19, 36].

An example of EHNBS construction is the high efficiency land/water ecotones at limited space (5 m strip) which contain denitrification barriers, plant buffer zone and possibility to incorporate geochemical barrier for phosphorus trapping [34] (**Figure 4.III.I**).

The large scale of harmonisation hydroengineering with EH NBS by constructing reservoir on the floodplain and maintaining the river continuum is introduced at **Figure 4.III.J**. The key assumption was to guarantee good water quality - no toxic algal blooms for the recreational reservoir while still maintaining the river continuum. Therefore it was necessary to consider as a reference point the long term analysis of river pulses for identification of periods with good water quality to use for supply of reservoir in water low in suspended matter and phosphorus concentration [7].

Such concept of multifunctional reservoir based on three principles of ecohydrology improves: W - Water resources by increase the amount of water retained in the river valley and its quality by enhancement of biological self-purification process in biofiltration system and diversity of habitats; B - biodiversity by enhancement diversity of aquatic and wetlands habitats; S – services for society – bathing and fishing; R – resilience to climate of all river valley ecosystems by increase of retentiveness of river valley and ground water reserve; CE - culture and education by building the education centre for teaching on importance of river valley in cultural development in history and to develop citizens science.

6. Ecohydrology: summary and way forward

Water is key driver of the strategy to reverse the degradation of Biosphere Sustainability (Sustainable Development Goals, SDG of UN), especially in the face of increasing climate changes. The fundamental step in establishing holistic strategies and systemic solutions should be to understand the hydrological mezocycle, water/biota interplay and hierarchy of drivers. All of that enforced by knowledge and broad scope of environmental sciences with consideration of diverse economic, legal and societal interactions. Therefore for engineering harmony between environment and society, there is **a need to develop an integrative sustainability science**. The fundamental step for achieving this is translating accumulated scientific information into knowledge (understanding the processes, feedback and the hierarchy of regulatory mechanisms in these systems), and then to translate this knowledge into wisdom: the ability to solve sustainability problems by innovative NBS and technologies integrated in systemic solutions which are mitigating human impacts and also increase adaptive capacity and strengthen the ability to adapt society and professional skills to new Evolutionary/Ecosystemic paradigm and relevant technologies, thus changing the hierarchy of needs and situations.

For acceleration of achieving biosphere sustainability (SDG UN) the further steps based on Ecohydrology are necessary:

1. **Proactive Education of society** - coping with uncertainty in changing world and especially climate. There is an urgent need for shaping societal attitudes and understanding of Sustainability: **understanding the biosphere as a dynamic system where water, carbon, phosphorus and nitrogen cycling**

serve as the primary drivers of sustainability, and that these are influenced by each of us and in turn determine the amount and quality of the ecosystem services we need. This approach requires proactive education, which will stimulate a shift from a Sociocentric/Mechanistic to an Evolutionary/Ecosystemic paradigm based on three assumptions: 1/the unity of Man and Nature, 2/ consciousness that happiness is not correlated with consumption but first and foremost, a fair and good relationship with other people, and with functioning in a healthy environment, 3/our positive mental status is to a great extent, defined by the quality of our environment. Such beliefs can to a great extent stimulate social capital – confidence and cooperation.

2. **Social capital as an important factor for translation of knowledge into wisdom and of innovations in catchment management**. The philosophy of the exchange of ideas and openness for controversial opinions has been broadening the holistic perception of the problems to be solved and generate most efficiently innovations. It is worth to underline that there appear to be many examples that both encouragement of team spirit but also respect for seniority and leadership towards achievement of strategic goals first and foremost reversing degradation of biosphere, synergies of both should also amplify the translation of Science into Technology.

3. **Socio-economic Foresight of the catchment as a tool for creating a desirable future**. Action without vision and strategies have typically resulted in a waste of human potential and resources. Hence, the primary tool used for the development of responsible vision and strategy in achieving SDG UN should be the foresight methodology, which should consider the circular economy, i.e. reduction of impact and bioeconomy production of commodities from renewable resources, and the enhancement of sustainability potential WBSRCE, where the water mezocycle has to be used as a framework for assessment, planning and management.

Author details

Maciej Zalewski[1,2]

1 European Regional Centre for Ecohydrology PAS, Łódź, Poland

2 UNESCO Chair of Ecohydrology and Applied Ecology University of Lodz, Łódź, Poland

*Address all correspondence to: mzal@biol.uni.lodz.pl

References

[1] Columbus Declaration. Harmonization of Societal Needs with the Eco Sphere in the Anthropocene Era. EcoSummit 2012. Columbus, Ohio, USA.

[2] Ecological Footprint. Global Footprint Network, available from: https://www.footprintnetwork.org/our-work/ecological-footprint/

[3] Naiman R.J., Decamps H., editors. The ecology and management of aquatic-terrestrial ecotons, UNESCO MAB series. Paris: The Parthenon Publishing Group/UNESCO. 1990.

[4] Naiman R.J., Decamps H., Fourier F, editors. Role of Land/Inland Water Ecotons in Landscape Management and Restoration. Vendome: Proposals for Collaborative Research, UNESCO 1989.

[5] Ryszkowski L., Kędziora A. Energy control of matter fluxes through land-water ecotones in an agricultural landscape, „Hydrobiologia" 1993; 251: 239-248.

[6] Biosphere II, University of Arizona, available from: https://biosphere2.org/

[7] Zalewski M. Ekohydrologia. Warszawa PWN. 2020. 262 p.

[8] Zalewski M., Robarts R. Ecohydrology - a new Paradigm for Integrated Water Resources Management. SIL News 2003; 40:1-5.

[9] Zalewski M., Naiman R.J. The regulation of riverine fish communities by a continuum of abiotic-biotic factors. In: J.S. Alabaster editor. Habitat modification and freshwater fisheries. London. Butterworths Scientific 1985.

[10] Zalewski M., Frankiewicz P., Brewińska B. The factors limiting growth and survival of Brown trout, *Salmo trutta* m. fario L. introduced to different types of streams, „Journal of Fish Biology" 1985; 27(A): 59-73.

[11] Krebs C. J. Ecology: The exoerimental analysis of distribution and abundance. New York: Harper and Row 1972.

[12] Southwood T.R.E.. Habitat, the templet for ecological strategies? Journal of Animal Ecology 1977; 46: 337-365.

[13] Zalewski M., Thorpe J.E., Scheimer F. Guest Editorial. Ecohydrology and Hydrobiology 2001; 1: 5-7.

[14] Zalewski M., Brewińska-Zaraś B., Frankiewicz P. Fry communities as a biomanipulating tool in a temperate lowland reservoir. Archiv für Hydrobiologie–BeiheftErgebnisse der Limnologie 1990; 33: 763-774.

[15] Bunn S.E., Arthington A.H. Basic principles and ecological consequences of altered flow regimes for aquatic biodiversity. Environmental Management 2002; 30: 492-507.

[16] Mitsch W., Jørgensen S.E. Ecological engineering and ecosystem restoration. New York: John Wiley & Sons 1989.

[17] Staskraba M., Tundisi J.G., editors. Theoretical Reservoir Ecology and Applications. Leiden, Backhuys 1999.

[18] Zalewski M. Fish Diversity and ecotonal habitat. In: J.-B. Lachavanne, J. Raphaëlle, editors. Biodiversity in land-inland water ecotones. UNESCO, Paris: The Parthenon Publishing Group Internationa Publishers in Science, Technology and Education. 1997; 183-203.

[19] Zalewski M. Ecohydrology-the scientific background to use ecosystem

properties as management tools toward sustainability of water resources. Ecological Engineering. 2000; 16: 1-8.

[20] Zalewski M. Ecohydrology and hydrologic engineering: regulation of hydrology-biota interactions for sustainability. Journal of Hydrological Engineering. 2014; 20 (1):A4014012

[21] Eagleson, P. S. Ecohydrology: Darwinian expression of vegetation form and function. Cambridge, U.K: Cambridge University Press; 2002.

[22] Rodriguez-Iturbe, I. Ecohydrology: A hydrological perspective of climate-soil-vegetation dynamic. Water Resources Research; 2000; 36(1): 3-9.

[23] Emaus, D., Hatton, T., Cook, P., and Colvin, Ch. Ecohydrology—Vegetation function, water and resource management. Collingwood, Australia; Commonwealth Scientific and Industrial Research Organization; 2006.

[24] Guswa A. J., Tetzlaff D., Selker J. S., Carlyle-Moses D. E., Boyer E. W., Bruen M., Cayuela C., Creed I. F., van de Giesen N., Grasso D., Hannah D. M., Hudson J. E., Hudson S. A., Iida S., Jackson R. B., Katul G. G., Kumagai T., Llorens P., Ribeiro F. L., Michalzik B., Nanko K., Oster C., Pataki D. E., Peters C. A., Rinaldo A., Sanchez Carretero D., Trifunovic B., Zalewski M., Haagsma M., Levia D. F. Advancing ecohydrology in the 21st century: A convergence of opportunities. Ecohydrology. 2020; 13:e2208

[25] Zalewski M., editor. Guidelines for the integrated management of the watershed – phytotechnology and ecohydrology, United Nations Environemant Programme, International Environmental Technology Centre. Freshwater Management Series. 2002; 5: 188.

[26] Wolanski, E., Estuarine Ecohydrology. Amsterdam: Elsevier. 2007; 157 pp.

[27] Chícharo, L., Chícharo, M.A. Applying the ecohydrology approach to the Guadiana estuary and coastal areas: lessons learned from dam impacted ecosystems. Estuarine Coastal and Shelf Science. 2006; 70: 1-2.

[28] Kiedrzyńska E., Jóźwik A., Kiedrzyński M., Zalewski M. Hierarchy of factors exerting an impact on the nutrient load of the Baltic Sea and sustainable management of its drainage basin. Marine Pollution Bulletin. 2014; 88:162-173.

[29] Kiedrzyńska E., Kiedrzyński M., Urbaniak M., Magnuszewski A., Skłodowski M., Wyrwicka A., Zalewski M. Point sources of nutrient pollution in the lowland river catchment in the context of the Baltic Sea eutrophication. Ecological Engineering. 20017; 70: 337-348.

[30] Chícharo M. A., Chícharo L. RNA:DNA Ratio and Other Nucleic Acid Derived Indices in Marine Ecology. International Journal of Molecular Sciences. 2008; 9: 1453-1471

[31] Wojtal-Frankiewicz A., Kruk A., Frankiewicz P., Oleksińska Z., Izydorczyk K. Longterm patterns in the population dynamics of *Daphnia longispina*, *Leptodora kindtii* and Cyanobacteria in a shallow reservoir: A Self-Organising Map (SOM) approach, PLoS ONE. 2015; 10 (12): e0144109. DOI: 10.1371/journal.pone.0144109.

[32] Mankiewicz J., Komárková J., Izydorczyk K., Jurczak T., Tarczyńska M., Zalewski M. Hepatotoxic Cyanobacterial Blooms in the Lakes of Northern Poland. Environmental Toxicology. 2005; 20(5):499-506

[33] Bednarek A., Stolarska M., Ubraniak M., Zalewski M. (2010), Application of permeable reactive barrier for reduction of nitrogen load in the agricultural areas – preliminary

results. International Journal of Ecohydrology and Hydrobiology. 2012 10, 2-4, 355-362.

[34] Izydorczyk K., Frątczak W., Drobniewska A., Cichowicz E., Michalska-Hejduk D., Gross R., Zalewski M. A biogeochemical barrier to enhance a buffer zone for reducing diffuse phosphorus pollution – preliminary results. Ecohydrology and Hydrobiology. 2013; 13(2): 104-112 DOI:10.1016/j.ecohyd.2013.06.003 .

[35] Kiedrzyńska E., Urbaniak M., Kiedrzyński M., Jóźwik A., Bednarek A., Gągała I., Zalewski M. The use of a hybrid Sequential Biofiltration System for the improvement of nutrient removal and PCB control in municipal wastewater. Scientific Reports. 2017; 7: 5477.

[36] Jurczak T., Wagner I., Kaczkowski Z., Szklarek S., Zalewski M. (2018), Hybrid system for the purification of street stormwater runoff supplying urban recreation reservoirs. Ecological Engineering. 2018; 110: 67-77.

[37] Krauze K., Wagner I. From classical water-ecosystem theories to nature-based solutions — Contextualizing nature-based solutions for sustainable city. Science of the Total Environment. 2019; 655: 697-706

[38] UN WWDR. World Water Development Report 2018 - Nature Based Solutions for water. Paris: UNESCO 2018.

[39] Harper D.M., Pacini N., Upton C., Morrison E.H.J., Fox R., Kiminta E. Water Cooperation for Sustainable Utilization: Lake Naivasha, Kenya. In: Free Flow – Reaching water security through cooperation. Paris: Tudor Rose for UNESCO. 2013.

[40] Zalewski M., Urbaniak M., Zerihun Negussie Y. Ecohydrological systematic solutions for reduction of siltation, eutrophication and dioxin-induced toxicity. The pilot study of the Asella BioFarm Park lake, Ethiopia. Ecohydrology and Hydrobiology. 2010; 10(2-4): 363-368.

[41] Negussie Y.Z., Urbaniak M., Szklarek S., Lont K., Gągała I., Zalewski M. Efficiency analysis of two sequential biofiltration systems in Poland and Ethiopia – the pilot study. Ecohydrology and Hydrobiology. 2012: 12(4): 271-285.

[42] Hehanussa P. E. Ecohydrology, a low cost hydh technology opportunity for Asian countries. Ecohydrology and Hydrobiology. 2002; 2 (1-4): 89-92.

[43] Taffarello D., Srinivasan R., Mohor G. S., Bittencourt Guimarães J. L., do Carmo Calijuri M., Mendiondo E. M. Modeling freshwater quality scenarios with ecosystem-based adaptation in the headwaters of the Cantareira system, Brazil. Hydrology and Earth Systems Sciences. 2018; 22: 4699-4723.

[44] Hobbs R.J. Novel ecosystems: theoretical and management aspects of the new ecological ward order. Global Ecological Biogeography. 2006; 15(1): 1-7.

[45] Kiedrzyńska E., Kiedrzyński M., Zalewski M. Flood sediment deposition and phosphorus retention in a lowland river floodplain: impact on water quality of a reservoir, Sulejów, Poland. Ecohydrology and Hydrobiology. 2008; 8(2-4): 281-289.

[46] Zalewski M. Ecohydrology – an interdisciplinary tool for integrated protection and management of water bodies. Archive of Hydrobiology. 2006; 158 (4): 613-622.

[47] Acreman M. C., editor. Hydro-Ecology linking hydrology and aquatic ecology. International Association of Hydrological Sciences. 2001. 162 p.

Chapter 5

Interlinking of River: Issues and Challenges

Pawan Jeet, Alok Kumar and Prem K. Sundaram

Abstract

Climate change events cause erratic spatial and temporal variability in rainfall, temperature, humidity, etc. in long term, and are most severely affecting irrigation, domestic and industrial water supply. At the same time, water availability is also under pressure due to climate change and overexploitation of water resources. In a monsoonal climate that is already erratic and highly seasonal in nature, this increased variability due to climate change will further impact water availability and salt water intrusion. To overcome such problems, one of the most effective ways is interlinking of rivers. It is the interbasin water transfer from the water surplus rivers to water deficit rivers or regions. It will increase water supply, irrigation potential, mitigate floods, and droughts and reduce regional imbalance in the availability of water. Interlinking of rivers will reduce regional imbalances significantly and provide benefits by the way of additional irrigation potential, domestic and industrial water supply, hydropower generation, and transport facilities.

Keywords: rainfall, river interlinking, surface water, groundwater, droughts, floods

1. Introduction

The rainfall occurrence in India is mainly due to orographic effect, coupled with tropical depressions originating in the Arabian Sea and the Bay of Bengal. It accounts for about 85% of the total rainfall in the country. The uncertainty in rainfall occurrence is a serious problem for the country marked by extended dry spells and fluctuations in seasonal and annual rainfall pattern. Most parts of the country are facing deficit rainfall trend and are subjected to large variations resulting in frequent droughts and floods conditions. Floods cause immense hardship to the population and enormous loss to the country. In the summer seasons as the rivers dry up and the ground water level goes down, the water availability for agricultural, industrial and drinking purposes becomes critical. While, some parts of the country do not have enough water even for raising a single crop due to regional rainfall alteration, on the other hand, surplus rainfall occurs in some parts of the country causing floods or waterlogged situation.

Irrigation has been the prime factor for increasing the food grain production in India from a mere 50 million tons in 1950s to more than 291 million tons in 2019–2020. In 1950–1951, the canal irrigated area was 8.3 million hectares while in 2013–2014 at 17 million hectares. Despite that, the relative importance of canals has fallen from 40% in 1951 to 26% in 2010–2011. On the other hand, in 1950–1951, the well and tube well accounted for 29% of gross irrigated area and they share 64% of the gross irrigated area in 2012–2013. Irrigated area has increased from 22 million hectares to 66 million hectares in the year 2012–2013 [1]. The population of India, which is around 1.2 billion at present, is expected to increase to 1.5 to 1.8 billion in the year 2050 and that would require about 450 million tons of food grains. The irrigation potential has to be increased to 160 million hectares for all crops for meeting food requirements by 2050. Through conventional sources, the maximum irrigation potential that could be created is about 140 million hectares. Other strategies shall have to be evolved for attaining a potential of 160 million hectares.

The Brahmaputra and Ganga rivers are the main Indian rivers in which almost 60 per cent of the Indian river drain. They also cause recurring floods and hence damages. Flood damages, which were Rs. 520 million in 1953, have gone up to Rs. 957.36 billion in 2018 mostly affecting the States of Assam, Bihar, Uttar Pradesh and West Bengal along with loss of lives. In other side large areas in the States of Andhra Pradesh, Gujarat, Rajasthan, Karnataka and Tamil Nadu face recurring droughts with approximately 85% areas of these states falls under drought prone.

One of the most effective means to enlarge the irrigation potential of river command areas is the Inter Basin Water Transfer (IBWT). It refers to the water transfer from water surplus rivers to the water deficit rivers or regions. Brahmaputra and Ganga rivers particularly their northern tributaries, Godavari, Mahanadi and West Flowing rivers originating from the Western Ghats of India are found to be surplus in surface water resources. If storage reservoirs can be built on these rivers and connected to other parts of the country, regional water imbalances could be reduced and many benefits by way of additional irrigation potential created, industrial and domestic water supply, hydropower generation, waterways facilities, etc. would be ensured.

2. History behind interlinking of rivers

The initial plan to interlink India's rivers came in 1858 from a British irrigation engineer, Sir Arthur Thomas Cotton, but the idea of interlinking Indian rivers was revived a few decades ago independently by M. Visveswarayya, K. L. Rao and D. J. Dastur. In 2002, the Supreme Court of India ordered the Indian Government to complete river interlinking project within the next 12–15 years. In response to this order, the Government of India appointed a Task Force and scientists, engineers, ecologists, biologists and policy makers started to deliberate over the technical, economic and eco-friendly feasibility of this gigantic project [2].

Since 2015, Indian Government has implemented river interlinking projects in several segments such as the Godavari-Krishna river interlining in Andhra Pradesh and the Ken-Betwa rivers interlink in Madhya Pradesh. These projects are built with aims that it will enhance annual per capita water availability for increasing population of the country. The Godavari-Krishna rivers interlinking projects also envisions an area more than twice the size of Andhra Pradesh receiving extra water for irrigation and to even out the unwarranted swings between droughts and floods. Yet even as the project moves forward there is a large possibility that it could dislocate nearly 1.5 million countries population due to the flooding of 27.66 lakh hectares of land [3].

3. Proposals for interlinking of rivers (ILR) in India

3.1 Earlier proposals

Transferring surplus water available in some regions to water deficit areas have been made suggested through formation of a National Water Grid from time to time. The two such proposals which attracted considerable attention put forth in the 1970s were:

3.1.1 National Water Grid by Dr. K.L. Rao (1972)

In 1972, Dr. K.L. Rao presents a 2640 km long Ganga—Cauvery river interlinking project. Its major component includes a large scale pumping of water over a head of 550 m. The power requirement for water lifting from the head was estimated to be 5000–7000 MW, used for irrigating an additional culturable area of 4 million hectares. This project had not any flood control benefits. It had estimated that the overall cost of the project was about Rs. 12,500 crores.

3.1.2 Garland Canal by Capt. Dastur (1977)

In 1977, Capt. Dastur presented a proposal for construction of two canals in Himalayan regions of India. The first proposed canal was 4200 km long Himalayan Canal run along the foothills of the Himalayas from the Ravi river to the Brahmaputra river near Chittagong. The second proposed canal is 9300 km long Garland Canal covering the Central Deccan and Southern Plateau region of the country. Both of the canals integrated with numerous lakes/reservoirs and interconnected with water pipelines at two points such as Delhi and Patna. The total estimated cost of the projects was Rs. 24,095 crores.

3.2 Existing proposals

Many large scale water transfer projects have been planned and few of them implemented and constructed as a landmark for the overall development of the water scarce regions. A few successfully implemented projects are briefly explained here.

- Periyar river project

The project was started in 1895 with the aim to provide irrigation facility to water deficit Vaigai river basin. This river project is one of the most outstanding endeavors of the nineteenth century in trans-basin diversion. The project envisages the transfer of water from Periyar river basin to Vaigai river basin. A masonry gravity dam at Periyar river has been constructed across a gorge on west flowing. Its height is about 47.28 m. A 1740 m long tunnel across the mountain with a discharge capacity of 40.75 cumecs has been driven to supply the water to Vaigai river basin. Initially, it provided irrigation to 57,923 ha culturable land, which has been extended to 81,069 ha. There is also a hydropower station of 140 MW capacities.

- Parambikulam Aliyar river project

It is an interstate multipurpose project completed in late 1960s and functioning based on an agreement between the states of Tamil Nadu and Kerala. Nine dams and two weirs had been constructed and their reservoirs interlinked by tunnels. The project envisages transfer of water from Chelakudi basin to Bharatapuzha and Cauvery basins. This project transfer water from the basins of three west flowing rivers originating from the western ghats of India along the Kerala-Tamil Nadu border such as Bharathapuzha Chalakkudipuzha Periyar river. These rivers are mainly dependent on the southwest monsoon and northeast monsoon rainfall. The water released to the east is mainly used for irrigation purpose. The water is being delivered to drought prone areas in Coimbatore district of Tamil Nadu and the Chittur area of Kerala states. The gross command area for irrigation is about 1,62,000 ha. There are four hydropower stations with an overall capacity of 185 MW.

- Kurnool Cudappah Canal project

A private company started this scheme in 1863. The project envisages transfer of water from Krishna basin to Pennar basin.

It takes off from right flank of Anicut constructed across Tungabhadra River near Sunkesula Village in Kurnool District. The total length of canal is 306 km i.e. from Sunkesula Anicut up to 235 km in Kurnool district and the remaining length of 71 km in Cuddapah district. The canal has total storage capacity of 84.9 cumecs which extends from Krishna to Pennar basin and irrigates area of about 52,746 ha.

- Telugu Ganga river Project

This project has been implemented primarily to meet the pressing need of water supply to Chennai metropolitan area as well as to irrigate 5.75 lakh acres in drought prone areas of Rayalaseema and uplands of Nellore District in Andhra Pradesh. It brings Krishna water from Srisailam reservoir through an open canal, first to Somasila reservoir in Pennar valley. The scheme consists of 408 km long canal from Srisailam Reservoir to Andhra Pradesh. From Somasila, water is capture to Kandaleru through a 45 km long canal and then to Poondi reservoir in Tamil Nadu through another 200 km long constructed canal. By mutual agreement, 12 TMC of water is to be delivered to Tamil Nadu at the border from Krishna basin. This greatly augments the water supply to Chennai city. The canal also irrigates 2.33 lakh hectares in Andhra Pradesh.

- Ravi-Beas-Sutlej-Indira Gandhi Nahar Project

This project presents an excellent example of the way the big inter basin water transfers initiatives added all round socio-economic improvement with typical enhancement inside the ecological and environmental factors of the vicinity. As per the Indus Water Treaty (1960) water of three eastern rivers viz. Beas, Sutlej and Ravi have been issued to India. As the land to be benefited from this interlinking river project in India, lies mainly to the east and south of these rivers basins, the rivers needed to be interlinked and the water transfer to canal systems for serving large tracts in India. Bhakra garage is the principal water garage on Sutlej river while Pong garage is the principal water garage on Beas river. Bhakra basin system provides irrigation to about 26.3 lakh hectares of new culturable area except stabilization of present irrigation facility to 9 lakh hectares. The gross hydropower generation capacity of Bhakra Nangal Project is 1379 MW. A Pondoh diversion dam is situated 140 km upstream of Pong reservoir on Beas which transfer water from Beas to Bhakra reservoir and generates hydropower of 165 MW. The Beas-Sutlej link is 37.25 km long of which 25.45 km is in tunnel through difficult hard rock formations. The overall discharge potential of the tunnel is 254.70 cumecs. Ranjit Sagar dam is also constructed at Ravi river that gives additional water to Beas and additionally generate a big block of hydropower.

4. Inter Basin water transfers in other countries

Many large-scale water transfer schemes have been planned and implemented in other countries also.

- South-north water transfer project, China

An ambitious plan to link Yangtze river basin in the south with the yellow river basin in the north, construction of the South-north water transfer project (SNWTP) began on 2002. Two third of the country's water is in the south, while half its people and nearly 65% of its agricultural land are in north. The project is set to cost nearly $80 billion and has necessitated the relocation of 330,000 people. He will transfer 45 billion cubic meter (BCM) of water through 3000 km long tunnels and canals.

- Tagus-Segura transfer project, Spain

This project was completed in 1978 that connects four river basins Tagus, Jucar, Segura and Guadiana, to irrigate 1.7 lakh hectares and provide water to 76 municipalities in south eastern Spain. The project has resulted in reduced in flows in Tagus.

- Lesotho highlands water projects, South Africa

This project was started in 1950 and completed in 1986 by South Africa and its neighbor Lesotho, the project involves transferring water from the upper reaches of the Orange river in Lesotho to the Vaal river in south Africa, and also generate hydel power. This project transferred 750 million cubic meter water (MCM) annually.

- California's State Water Project, United States

The 1st phase of the project was completed in the year 1973. The California State Water Project is a water storage and delivery system of reservoirs, aqueducts, power plants and pumping plants. The major purpose of the project was water supply. It diverts 4 km^3 of water from excess watered northern California to the drier central and southern parts. The conveyance system consists of 715 km long aqueduct, a complex system of lined and unlined canals, siphons, tunnels and pumping stations, etc. The water also irrigates about 750,000 acres of farmland, mainly in the San Joaquin Valley.

Similarly, importantly major under construction and existing inter basin water diverted in Canada include Kemano, Churchill Diversion, Well and Canal, Bay d' Espoir, James Bay, Churchill Falls, etc. Proposed river inter basin transfers in Canada include Long Lake, Ogoki (for transfer within Canada) and North American Water and Power Alliance (NAWAPA), Canadian Water, Grand Canal Concept, Central North American Water Project (CNAWP), Magnum Plan, Smith Plan etc. for transfer from Canada to USA.

5. Need for Inter Basin water transfers (IBWT)

Inter Basin Water Transfers is necessarily required to overcome the water scarcity situations in the regions/basins. These are needed to enhance water utility and reduce water wastage of water surplus areas in the following manner:

- Large variation in rainfall and available water resources in space and time
- Diversion of water from water surplus basins to water deficit basins/regions
- Use of the surplus water which is otherwise flowing into the sea unutilized
- To mitigate likely adverse impact of climate change, short term and long term

5.1 Indian National Water Policy (2012)

National water policy (NWP) considered water as economic goods for promoting its conservation and efficient use. NWP was formulated to govern the planning and development of water resources and their optimum utilization. The NWP was adopted in September, 1987. It was reviewed and updated in 2002 and later in 2012. It stated that Inter basin water transfers are not only for increasing production but also for meeting basic requirement of human need and achieving equity and social justice [4]. Inter-basin transfers of water should be considered after evaluating the environmental, economic and social impacts of such transfers.

6. Issues: interlinking of Indian rivers

In case of water disputes, Article 262 of the constitution of India provides "Parliament may by law provide for the adjudication of any dispute or complaint with respect to the use, distribution or control of the waters of, or in, any inter-State river or river valley."

According to the National Water Policy (NWP), water is a prime natural resource for human's beings as well as animals and, for this reason, a valuable countrywide asset. These days, it is difficult to explore freshwater due to boom in population, agricultural and industrial sectors, and contamination of water assets. By 2020, the world human population is expected to reach up to 7.9 billion and the world may face to great severity for freshwater. The India's annually receives about 4000 km^3 of water from precipitation, but due to exclusive rainfall patterns and their mismanagement, leads to wastage of water. By the way of considering this interlinking of Indian rivers, a highly formidable and big project is planned, that is under debate.

In 1990, the Government of India appointed a high level Commission to have a look at the approach of water resource development in India, together with the possibility of interlinking rivers. The concept of canals linking is to divert surplus waters from some identified rivers basins to the water deficit river basins and regions [5].

In 2015, the Supreme Court of India directed the Government of India to draw up and put in force a program to interlink principal rivers of India. Subsequently, the Prime Minister of India announced its decision to work on the court directive and appointed a task force to ensure the implementation of the project by the same year [6].

The appeal of interlinking rivers is primarily based on the expertise that a huge quantity of water of the country rivers flows into the ocean and that if only this is prevented, and water transferred from water surplus rivers to water deficit rivers, there may be good enough supply of water for anyone throughout the country. At another level, the project is seen as promoting national integration and a fair & equitable sharing of the country's water wealth. Whether or not the linking of rivers will promote integration or generate greater disputes is a moot question. The expert belief that the river interlinking is essential to make sure adequate and safe water delivery to all people and anywhere. Domestic water use currently accounts for about 5% of the total water use extracted through wells, tube-wells, tanks and canals.

The requirements of water in different sectors growing rapidly but will still are relatively small compared to those of other uses. To meet the requirement under limited conditions, river interlinking is the only solution for this problem. Even if interlinking were justified for other reasons, it will be difficult to reach water to the entire livelihood without huge investments in a centralized water distribution network. Through approaches of decentralized local rainwater harvesting, by renovating and improving traditional water storage structure can meet the essential demands of water for domestic requirement more effectively and at a low input cost.

By far, agriculture is the largest water user and more than 85% of water from wells, tube-wells, tanks and canals are used for irrigation purposes. The water demand is increasing and will continue to be, by far, the biggest claimant on available water supplies. There is much scope for enhancing the water use efficiency of irrigation systems by reducing wastage of water such as runoff, seepage, evaporation, interflow and so on, and through efficient on-farm water management practices.

The water demand for irrigation arises when rainfall is not sufficient to meet the water demand for raising crops and obtaining optimum crop yields. The annual rainfall is sufficient to meet the water requirements in the *kharif* in the country. Irrigation is required essentially to tide over insufficient soil moisture during dry spells within the season. In the states Haryana, parts of Gujarat, Punjab, Rajasthan and Tamil Nadu agricultural acreage which required irrigation during the *kharif*. Practically in the whole country, especially the northwest, irrigation is essential between the month of November and June. So far, these water imbalances have been met by water capturing technologies such as constructing water harvesting structures to store excess rainy water during monsoon for use in the dry season and by exploring groundwater. Some areas, such as Tamil Nadu, have exhausted the water potential for harnessing the surface flows. In several other part of the country, the possibilities for constructing water storage structure are limited. In these areas groundwater resources are already overexploiting to meet the water demands of different sector, so under these conditions the scope for exploring water is limited. In many areas, the problem is to check expansion and contain the rate of exploitation of surface and sub-surface water resources. Considering this river interlinking is seen as a way out to solve the problem of water deficit of the country.

An examination of the river interlinking raises numerous questions: (1) it is primarily based at the presumption that there are large surplus flows in some basins and that the physical transfer is possible in terms of physical engineering, and can be performed economically without growing any adverse environmental, social and economic impact.

It is the fact that the most of the difficulty arises practically in all Indian rivers during the southwest monsoon. About 90% of south Indian rivers flow takes place in the month of May to November. Being perennial Indo-Gangetic and Brahmaputra river basins, the percentage of the total flow occurring during months between May and November may be somewhat smaller but not all that much smaller flow rate. More than 80% of the yearly flow in the Kosi river occurs in the month between May and November and almost three-fourth in the month between June and October.

The prevalence of monsoon refers to the season when aggregate rainfall is sufficient for plant growth. In some parts of India, such as Rajasthan, parts of Gujarat and the Deccan, even the *kharif* rain is far too low and variable for productive agriculture. In some others parts, excessive water could assist transfer to more productive cropping patterns. Those deficit regions are far from those taken into consideration surplus water are required to transfer over very water harsh terrain.

7. Hashim commission report (2004–2005)

The commission report highlighted that which rivers and at which locations water surpluses could be transfer and to which rivers, and at what factors in these rivers the transfer water could be taken. There may be no records at the large quantity of water to be transferred via different canals link, the location and extent of the region to be strengthen on the receiving side and the water distribution system via which water is to be allocated in the region.

8. Interlinking of rivers to get new push during the year 1999–2004

An ambitious project on interlinking of river got new impetus during the year 1999–2004. It focuses mainly on rivers linking of sub-basins within a larger basin or nearby basins instead of going on for distant inter-basin river linkages.

Government of India considered those rivers basin for interlinking which are adjacent to each other, keeping in the mind its feasibility and utility to larger beneficiaries/stakeholders. It has done the interlinking of rivers in a manner that it simultaneously looks after irrigation and drinking water needs of human beings and ecological concerns.

The International Water Management Institute (IWMI), Sri Lanka and the Challenge Program on Water and Food (CPWF) have designed a 3 year project on "Strategic Analysis of India's River Linking Project" to qualitatively improve the troubles and route of the prevailing NRLP debate [7].

9. Challenges: interlinking of Indian rivers

9.1 River linking should adjust rainfall, hit monsoons

Criticizing the interlinking of rivers (ILR) mission of the Government of India, a leading geologist and environmental expert warned the move ought to disrupt rainfall pattern which could be a major problem in respect of climate change. The formidable ILR initiative which received a boost via the Prime Minister of India has 30 river linking initiatives under its ambit and consists of each peninsular and Himalayan Rivers.

V. Rajamani (Emeritus Professor, Jawaharlal Nehru University, New Delhi) reported that there is a major disruption of ecosystem. In respect of climate change there is a possibility of change in rainfall pattern. The marine ecosystem will be disturbed and the physical process for the rainfall will be affected. You could no longer dependent upon the monsoon. Under changing rainfall pattern, it is not possible to meet water requirement of the ecosystem. So, there is need to link water bodies of those areas with water surplus areas to enhance the water flow to meet the water necessities.

There is not enough water to interlink rivers across India: IIT study.

A study of Indian Institutes of Technology, Mumbai and Indian Institutes of Technology, Chennai reported that rainfall of the country has decreased over the years 1901–2004, reducing water storage even in the river basins that have surplus water. It also reported that a significant fall in rainfall (i.e., more than 10% in each basin) in the major water surplus river basins of the Godavari, the Mahanadi, the Mahi, the Brahmani, the Meghna and the many small rivers in the Western Ghats, and east flowing river basins of the country. Only the Brahmaputra river basin showed that there is no deficit in rainfall [8]. Rivers linking project will have an ecological effect while building a network of dams, reservoirs and canals. It should be reanalyzed and reevaluated through considering changes in climatic patterns of the river basins. In such manner a decrease in surplus river basin contradicts the conventional perception that climate change phenomenon resulting wet areas to turn out to be wetter and dry areas to turn out to be drier over Indian conditions.

In 1982, National Water Development Agency (NWDA) was set up as an autonomous organization under the Ministry of Irrigation with the aim to carry out detailed studies, surveys and investigations of the water balance and feasibility studies of the river interlinking projects.

In 2002, the President of India point out the river linking project in the course of a speech. He proposed that it as a solution to India's water problem after which an application requesting an order from the Supreme Court of India on river linking matter was submitted. Finally, in 2002 the Government of India declared a substantial plan for an IBWT program involving 30 links of different river basins of peninsular and the Himalayan parts of India [9].

The interlinking of river (ILR) was introduced in 1982; it was actively taken up during Atal Bihari Vajpayee's tenure as Prime Minister during 1999–2004. The ILR project has two components the peninsular and the Himalayan. Both the components together have 30 river-linking projects.

The peninsular part of the India covers the rivers in southern India envisaged developing a "Southern Water Grid" with 16 important river linkages in different states. This part included diversion of the excess waters of Godavari and Mahanadi to the Krishna, Cauvery, Pennar and Vaigai rivers. The inter basin water transfer network in peninsular part of the India is shown in **Figure 1**.

The peninsular part of the India has 16 major canals and 4 sub-components: (1) Network of Mahanadi-Godavari-Krishna-Cauvery-Vaigai rivers; (2) Network of west flowing rivers lies between south of Tapi and north of Bombay; (3) Network of Parbati-Kalisindh-Chambal and Ken-Betwa rivers and (4) diverting the flow in some of the west flowing rivers to the eastern side of the country. As per NRLP the enroute irrigation under the peninsular part of the country is expected to irrigate substantial areas. The amount of water diverted in the peninsular part may be 141 km^3. The area to be irrigated is situated in arid and semi-arid western and peninsular part of India. The overall project cost includes three components in the peninsular part of the India may be Rs.1,06,000 crore and the hydroelectric power component may be Rs. 2,69,000 crore. The gross hydroelectric power generated may be 4 to 34 GW [10]. The Himalayan part of the India was conceived for building storage reservoirs on the Ganga and the Brahmaputra and their main tributaries

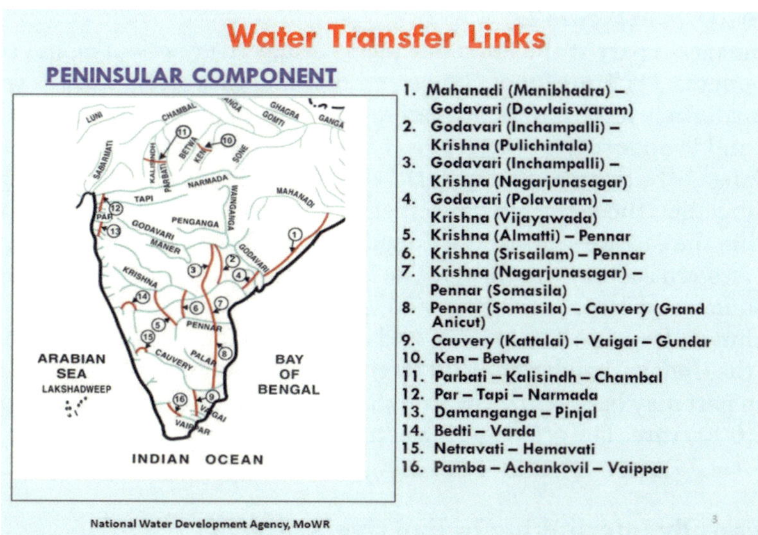

Figure 1.
Inter Basin water transfer network in peninsular India.

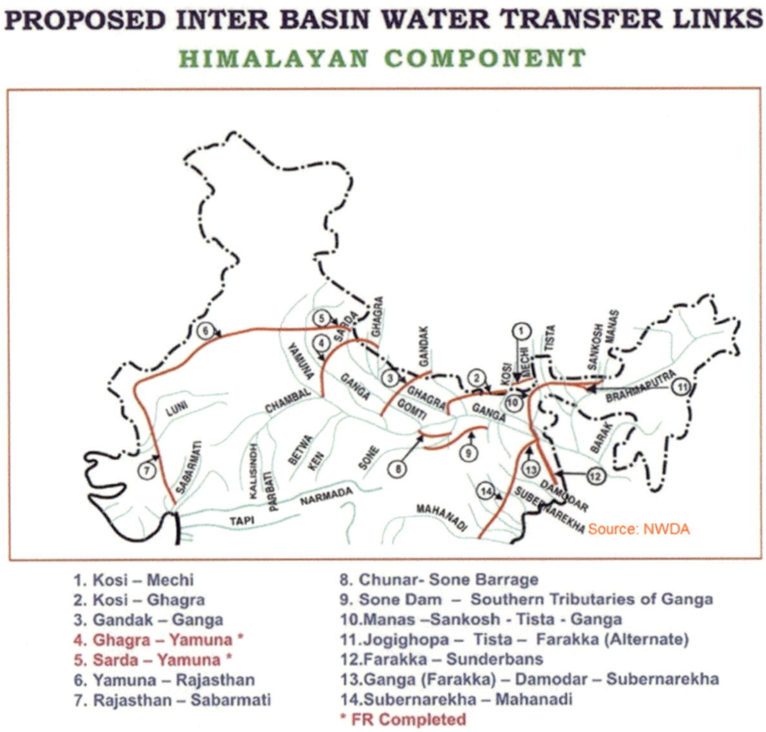

Figure 2.
Inter Basin water transfer network in Himalayan India.

both in India and Nepal in order to conserve water during the monsoon season for irrigation and hydro-power generation, besides checking floods. The Himalayan part of the India is comprised of 14 inter basin water transfer network including Brahmaputra-Ganga, Kosi-Mech, Kosi-Ghagra, Gandak-Ganga, Ghagra-Yamuna, Sarda-Yamuna, Ganga-Damodar-Subernarekha, Subernarekha-Mahanadi and Farakka-Sunderbans. The Inter basin water transfer network in Himalayan part of the India is shown in **Figure 2**.

The Himalayan part of the India has 16 important river networks, has two sub-components: (1) Transfer of Ganga and Brahmaputra rivers surplus waters to the Mahanadi Basin and from Mahanadi to Godavari, Godavari to Krishna, Krishna to Pennar and Pennar to the Cauvery river basins. (2) Transfer of water from the Eastern Ganga tributaries to the western sects of the Ganga and the Sabarmati river basins. Altogether, those river water transfers network will mitigate the floods issues within the eastern sects of the Ganga Basin and gives irrigation water delivery to the western sects of the Ganga. The Himalayan sects needed large number of dams in Bhutan and Nepal to capture and divert flood waters from the tributaries of the Brahmaputra and Ganga rivers, and also within India to divert the surplus waters of the Godavari and Mahanadi rivers. The amount of water diverted in the Himalayan part may be 33 km^3. The overall project cost of the Himalayan part is to be Rs.1,85,000 crore. The gross hydroelectric power generated may be 30 GW [11].

10. Successfully interlinking Indian rivers projects

There are many river linking projects are successfully completed or operated by the Govt. of India is explained below:

10.1 Polavaram project

It is also called Indira Sagar project. It is a multipurpose project built on Godavari river in the state of Andhra Pradesh. It has assessed culturable command area (CCA) of 2.91 lakh hectares and hydropower generation capacity of 960 Mega Watt (MW). It additionally has a carrying capacity of drinking water supply of 23.44 thousand Million Cubic Feet (TMC) to Vishakhapatnam Steel Plant. Its' inter-basin annual water carrying capacity is 80 TMC to Krishna river basin [12].

10.2 Ken-Betwa link project

- The main aim of Ken-Betwa link Project provides 6.35 lakh ha irrigation and 49 MCM drinking water supply in the drought prone and backward Bundelkhand region of Madhya Pradesh and Uttar Pradesh.
- It will transfer surplus water of Ken basin to water deficit Betwa basin besides power generation of 78 MW.
- Estimated Cost Rs. 17,700 Crores
- 77 m high, 2031 m long Daudhan Dam on Ken river in Chhatarpur district of M.P 221 km long link canal.

10.3 Damanganga-Pinjal link project

- The project envisages to provide 579 MCM of water of Damanganga basin for domestic and industrial water supply to Mumbai city
- Project benefits
    - Domestic and industrial water supply to Mumbai city—579 MCM
    - Hydropower—5 MCM
- Total estimated cost—Rs. 2746.61 crore
- Benefit cost ratio—1.95

10.4 Par-Tapi-Narmada link project

- The main aim of this project is to provide irrigation for 2.30 lakh hectares by transferring 1330 MCM of water from Par, Auranga, Ambika and Purna rivers to water short north Gujarat Kutch region besides enroute irrigation and hydropower generation 21 MW.
- Project benefits
    - Annual irrigation—2.3 lakh ha
    - Hydropower—21 MW
- Total cost estimated—Rs. 9279 crore
- Benefit cost ratio—1.95

10.5 Mahanadi-Godavari link project

- NWDA reported that Godavari and Mahanadi river basins are water surplus basins. The joint surpluses water of these basins after accounting in basin uses in closing stage of improvement can be transferred to fulfill the water requirement of water deficit basins in South upto river Gundar via Mahanadi-Godavari-Krishna-Pennar-Cauvery-Vaigai-Gundar river linkages. According to the Government of Odisha surveys, the proposed dam turned into having submergence of 59,400 ha.

- Government of Odisha has proposed a dam at Barmul 14 km upstream of Manibhadra village on Mahanadi river. The Barmul dam with full reservoir level (FRL) of 80 m, height of 25 m can have a total storage capacity of 1216 MCM and water diversion of 9182 MCM. Out of which, 4046 MCM water can be diverted to Godavari river.

- Projected Benefits

    - Annual irrigation—5.03 lakh ha.

    - Drinking water supply—125 MCM

    - Hydro power generation—240 MW

10.6 Manas-Sankosh-Teesta-Ganga (M-S-T-G) link project

Manas-Sankosh-Teesta-Ganga (MSTG) link is proposed under the Himalayan Component of national Perspective Plan (NPP). MSTG link canal envisages diversion of the surplus waters of Manas and Sankosh rivers to Ganga at Farakka and further transfer to water shortage areas of Krishna, Pennar and Cauvery basins and providing irrigation facilities to the enroute canal command areas.

10.7 Kosi Mechi link project

- The main aim of this project is to provide irrigation benefits to the water scarce Mahananda basin command in the districts of Araria, Kishanganj, Purnea and Katihar during kharif season depending upon the pondage available in Hanuman Nagar barrage.

- Projected Benefits

    - Annual irrigation: 2.15 lakh ha

- Benefit cost ratio—3.66

10.8 Burhi Gandak Noon Baya Ganga link project

- The main purpose of this project is to diversion of flood water from river Burhi Gandak to river Baya/Ganga for flood moderation. Irrigation to the tune of 1.26 lakh ha during kharif season in Samastipur, Begusarai and Khagaria districts.

- Projected Benefits
  - Annual irrigation—1.25 lakh ha
- Total estimated cost—4214 crore
- Benefit cost ratio—1.54

11. Conclusion

The interlinking of rivers project is a major challenge and an opportunity to deal with the water related problems springing up drought, floods, climate change and so on. The long term strategy to water deficit problem lies in making the interlinking of rivers challenges by building a network of dams, reservoir, barrage, hydropower structures and canals throughout the geographical regions of the country. However, Interlinking of rivers is definitely a good solution for the shortage of water, but interlinking has to take place after a reconnaissance survey and detailed study so that does not cause any trouble to the environment or aquatic life.

Author details

Pawan Jeet[1]*, Alok Kumar[2] and Prem K. Sundaram[1]

1 ICAR-Research Complex for Eastern Region, Patna, Bihar, India

2 ICAR-National Academy of Agricultural Research Management, Hyderabad, Telangana, India

*Address all correspondence to: pawan.btag@gmail.com

References

[1] Government of India. State of Indian Agriculture, 2015-16. New Delhi: Ministry of Agriculture and Farmers Welfare. 2016;1-252

[2] Ameer S, Rafi A, Ahmad A. Interlinking of rivers: Concerns and benefits. Journal of Civil Engineering and Environmental Technology. 2016;**3**(6):544-548. ISSN: 2349-8404; e-ISSN: 2349-879X

[3] Koshy J, Bansal S. Interlinking of Rivers: An Idea with Flaws. The Hindu News. 2016. Available from: https://www.thehindu.com/data/last-drop-interlinking-an-idea-with-flaws/article8567203.ece

[4] Ministry of Water Resources. National Water Policy-2012. 2012. Available from: http://mowr.gov.in/policies-guideline/policies/national-water-policy

[5] Bandyopadhyay J, Perveen S. Interlinking of Rivers in India: Assessing the justifications. Economic and Political Weekly. 2004;**39**(50):5307-5316. DOI: 10.2307/4415896

[6] Vaidyanathan A. Interlinking of Rivers-I. The Hindu News. 2003

[7] International Water Management Institute. World water supply and demand. Colombo, Sri Lanka: International Water Management Institute. 2000;1-29. ISBN:92-9090-400-3

[8] Fernandes S. There Isn't Enough Water to Interlink Rivers Across India: IIT study. Hindustan Times News. 2016. Available from: https://www.hindustantimes.com/mumbai-news/there-isn-t-enough-water-to-interlink-rivers-across-india-iit-study/storyN65MtpqHL0x0YmYeuhH25M.html

[9] Thatte CD. Inter-basin water transfer (IBWT) for the augmentation of water resources in India: A review of needs, plans, status and prospects. International Journal of Water Resources Development. *2008;23(4):709-725*. DOI: 10.1080/07900620701488646

[10] Mehta D, Mehta NK. Interlinking of rivers in India: Issues & challenges. Geo-Eco-Marina. 2013;**19**:137-143. ISSN: 1224-6808

[11] Rath N. Linking of rivers: Some elementary arithmetic. Economic and Political Weekly. 2003;**38**(29):3032-3033. DOI: 10.2307/4413796

[12] Press Trust of India. Bihar Launches Ambitious River Linking Projects. The Hindu News. 2011. Available from: https://www.thehindu.com/news/national/other-states/bihar-launches-ambitious-river-linking projects/article2073483.ece

Chapter 6

Rivers of Lebanon: Significant Water Resources under Threats

Amin Shaban

Abstract

Lebanon is known by tremendous water resources, and this has been often viewed from the considerable number of rivers (i.e. 14 rivers). These rivers are characterized by small catchments and short length. The estimated average annual discharge from these rivers is approximately 2800 million m^3. Due to the sloping terrain of Lebanon; however, it was estimated that more than 75% of water from rivers is unexploited it mainly outlets into the sea. The majority of water use from the Lebanese rivers implies domestic, agriculture, as well as some other rivers are used for hydro-power generation where they contribute by about 20% of electricity needed for Lebanon. Lately, and added to water pollution, there is abrupt decline in the discharge from these rivers estimated to more than 60% of their average annual discharge. This unfavorable situation is attributed, in addition to the changing climate, to the anthropogenic interference is the most affecting one and it is represented by over pumping from these rivers and form the recharge zone for groundwater and springs that feed these rivers. This chapter aims at introducing a discussion on the existed challenges on the Lebanese rivers and the proposed and their impact.

Keywords: stream flow, over pumping, climate change, pollution, dams

1. Introduction

Rivers in Lebanon are usually considered as the most significant water resources, as well as they are one of the known heritage sites and distinguished landscape. They show dense networks since there are 14 rivers located in the area of 10,452 km^2. Therefore, the estimated total annual discharge is about 3452 million m^3 where there is about 20% of this discharge goes into the Transboundary Rivers, therefore, the net discharge from the Lebanese rivers is approximately 2800 million m^3/year.

The average annual discharge from these rivers is about 247 million m^3, which is low enough if compared with international rivers, as an example, this volume of water is equivalent to the discharge from the Nile River in one day. There is also an argument about the exact number of rivers in Lebanon notably that not all of the 14 rivers, which are discharging water all year long as it was the case in the past, and this gives the first indicator about the current bad status on these rivers. Thus, the adoption of this number (i.e. 14) has been derived from the morphometric characterization basis and catchment shapes of the existing perennial watercourses, even though this conflicts with their hydrology where water flow is now almost intermittent.

There is remarkable hydrologic feature that characterizes the Lebanese rivers where all of them are almost controlled by the rugged topography, and then distributed within the three geomorphological units of Lebanon (i.e. Mount-Lebanon, Bekaa Plain and the Anti-Lebanon). Hence, Lebanon with its small area represents as a regional hydrologic junction where water flows into three regional directions. These are regional flows: 1) northward to comprise a tributary of the Orates River Flow System, 2) southward forming a major tributary for Jordan River Flow System and 3) eastward where the Lebanese Coastal Rivers System flow to the Mediterranean Sea [1].

Rivers in Lebanon have diverse orientations of distribution on terrain surfaces, and more specifically they are characterized by different flow directions and dimensions. Therefore, these rivers can be classified as follows:

1. Coastal Rivers: These are 10 rivers in Lebanon span along Mount-Lebanon where they have originated from, and then trending from east and discharge in the Mediterranean Sea (**Figure 1**), and then described as "Coastal rivers". These rivers are essentially fed from the snowmelt. The coastal rivers are relatively short where the longest one is El-Awali River which is about 61 km (curved). In addition, these rivers have nearly similar basin characteristics where the channel slope is relatively high and averages at 35–40 m/km, and this makes water flows rapidly between 5- and 10 km/hour in average [2]. This in turn results water loss into the sea.

2. Inner Rivers: Other than the 10 coastal rivers, there are 4 rivers in Lebanon that are characterized by diverse catchment morphometry and hydrology including mainly the flow direction and discharge regime. One of these rivers (i.e. El-Kabir River) is originated from the most northern part of Lebanon and outlets into the Mediterranean Sea (**Figure 1**); another two rivers (i.e. Al-Assi River and Litani River) are originated from the Bekaa Plain where the first one flows northward to Syria and the second flows southward and then diverted

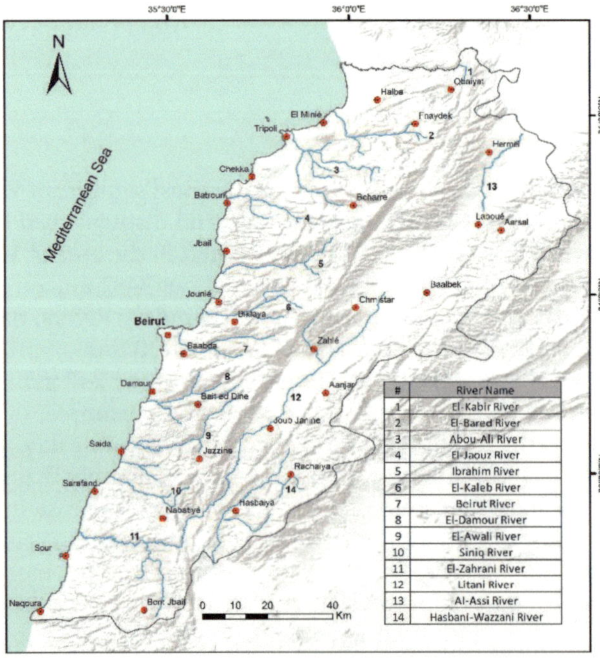

Figure 1.
Rivers of Lebanon [1].

towards the sea within the Lebanese territory. The fourth river (i.e. Hasbani-Wazzani River) is originated from Jabal Hermoun and then spans southward comprising the highest channel slope (40 m/km) of the four rivers. Except the Litani River, the other three rivers are Transboundary water resources.

2. Watersheds of the Lebanese Rivers

Elaborating the dimensions and mapping of drainage systems, including the catchment and the streams inside, is usually applied as a primary phase for detailed hydrological analysis and surface water assessment. Therefore, the geometric and morphometric analysis are utmost significant in watershed management to presume, for example, site suitability for surface water accumulation and harvesting, agricultural projects, dams' construction, hydro-power sites, etc.

Drainage systems of Lebanon including rivers were extracted directly from the stereoscopic satellite images (i.e. SRTM DEM) where digital elevation models were generated by magnifying the pixel details, and then slopes were extracted to determine flow directions and then stream delineation. Therefore, the digital extraction of drainage systems enabled calculating a number of geometric and morphometric measurements.

2.1 Geometry of the Lebanese rivers

Geometric measurements represent the calculations of the variables for the boundary of watershed (or catchment), and this will be totally separated from the properties of the streams (primary or secondary) included in the catchment. In this view, catchments with relatively large areas are usually subdivided into sub-catchments, which is dependent of the purpose of study applied.

For Lebanon, the area of rivers catchments is small and averaging about 250 km^2 for the coastal rivers if excluding the Litani River, which is an inner-coastal river. Hence, the largest area belongs to this river (i.e. Litani) and the smallest one belongs to Siniq River (**Figure 1** and **Table 1**).

1. Basin maximum length (B_l): This represents the maximum straight length of the catchment where it extends almost parallel to the primary watercourse, and it reflects the topographic orientation of a catchment

   B_L is a function of water arrival time to reach the outlet, and thus it controls the time of leakage, evaporation, and transpiration **Table 1** shows the maximum length of catchments of the Lebanese rivers

2. Basin width (B_w): The ratio of length of a catchment to its width significantly affects water flow to the outlet, thus when the difference between the length and width of a catchment is relatively low, that means the flow will be more regular and takes more time than it when this difference is high. For Lebanon, the B_w of rivers catchments are shown in **Table 1**.

3. Elongation Index (E_i): This represents the ratio between the diameter of the circle with the same area (A) as the catchment, and the distance between the maximum two points (B_l) in the catchment [3]. It is expressed by the formula:

$$E_i = \frac{2\sqrt{A}}{B_l\sqrt{\pi}} \tag{1}$$

No.	Catchment	A (km^2)	B_l (km)	B_w (km)	E_i	F_f	R_r
1	El-Kabir R.	303*	43	7.5	0.35	0.10	0.34
2	El-Bared R.	284	27	11	0.70	0.38	0.25
3	Abou-Ali R.	482	35	16	0.69	0.38	0.46
4	El-Jaouz R.	196	32	6.5	0.49	0.20	0.42
5	Ibrahim R.	326	40	8.5	0.57	0.20	0.47
6	El-Kaleb R.	237	36	12	0.48	0.18	0.57
7	Beirut R.	216	31	9	0.44	0.22	0.53
8	Ed-Damour R.	333	32	10	0.65	036	0.51
9	El-Awali R. R.	291	33	9.5	0.58	0.27	0.33
10	Siniq R.	102	19	5.5	0.60	0.28	0.26
11	El-Zahrani R.	140	28	6.5	0.48	0.18	0.28
12	Litani R.	2110	145	16	0.36	0.10	0.21
13	Al-Assi R.	1930*	51	31	0.98	0.76	0.25
14	Hasbani-Wazzani R.	645*	52	11	0.55	0.24	0.27

**Catchment area within Lebanon.*
R = River.

Table 1.
Geometric measurements of the Lebanese rivers' catchments [1].

Table 1 shows the calculated E_i for the catchments of the Lebanese rivers. According to Schumm [3], E_i is: < 0.5, 0.5–0.7, 0.7–0.8, 0.8–0.9 and 0.9–1 for more elongated, elongated, less elongated, oval and circular; respectively.

4. Form Factor (F_f): It is the numerical index used to determine ratio of the basin area to square of the basin length [1]. It is a function of the flow energy in the catchment. Thus, form factors for the catchments of the Lebanese rivers are shown in **Table 1**.

Hence, F_f must be less than 0.7854 [4]. Therefore, smaller F_f value indicates more elongated, while high F_f value experience larger peak flows of shorter duration. According to Horton (1932) form factor is expressed as:

$$F_f = \frac{A}{L^2} \tag{2}$$

5. Relief gradient (R_r): This is the ratio between the altitude at the highest and lowest points on the catchment, and it is calculated according to following formula [5]:

$$E = \frac{\text{Mean Elevation–Minimum Elevation}}{\text{Maximum Elevation–Minimum elevation}} \tag{3}$$

2.2 Morphometry of the Lebanese Rivers

These represent measurements for the dimensions, orientation and the connection between different streams in a catchment [6]. Thus stream morphometry evidences the origin and evolution of drainage networks, geomorphology and

No	Catchment	Length (km)		S_s	C_s	D_d	M_r	T_t
		Straight	Curved	m/km	—	S/km^2	%	—
1	El-Kabir R.	46	59	17	25	3.00	78	3.77
2	Al-Bared R.	37	49	13	14	1.00	76	2.75
3	Abou-Ali R.	42	54	54	46	1.75	78	6.95
4	Ej-Jouz R.	33	37	27	44	2.40	89	5.51
5	Ibrahim R.	44	50	63	45	5.30	88	16.30
6	El-Kaleb R.	35	41	66	45	6.10	85	17.25
7	Beirut R.	48	58	52	50	5.80	83	14.89
8	Ed-Damour R.	45	54	51	46	5.30	83	18.15
9	El-Awali R.	50	61	36	44	4.35	82	10.90
10	Siniq R.	18	21	7	9	2.35	86	7.54
11	Ez-Zahrani R.	36	41	8	13	2.10	88	8.16
12	Litani R.	163	174	5	8	0.84	83	0.54
13	Al-Assi R.	31	33	19	21	1.27	94	13.03
14	Hasbani-Wazzani R.	22	25	14	11	1.14	88	5.06

Table 2.
Major morphometric measurements of the Lebanese rivers' catchments [1].

geology of the underlying stratum. For this reason, stream morphometry controls water flow regime and mainly the flow energy. **Table 2** shows the main morphometric calculations for the streams (primary and secondary) in the Lebanese rivers' catchments.

1. Mean stream slope (S_s): It is the difference between the altitude at the source and the altitude at the outlet with respect to the total stream length. Thus, higher S_s results high flow rate along the primary stream in the catchment and vice versa. The following formula represents S_s [7]:

$$S_s = \frac{\text{Elevation at source–Elevation at outlet point}}{\text{Length of stream}} \quad (4)$$

2. Mean catchment slope (C_s): This is calculated by dividing the difference in elevation between points at defined lengths of the catchment (e.g. 0.85 L towards the upper and 0.10 L near the lower part of the catchment) over the length of the catchment. Hence, C_s is expressed by the following formula [8]:

$$S_b = \frac{\text{(Elevation at 0.85 L)–(Elevation at 0.10 L)}}{\text{Elevation at 0.75 L}} \quad (5)$$

3. Drainage density (D_d): This represents the degree of streams congestion in an identified area of the catchment. Thus, it is calculated by dividing the total length of streams within the identified area in the catchment. Hence, streams with high density indicate lower permeability of terrain surface if compared with lower density stream [9]. D_s is calculated according to the following formula:

$$D_d = \frac{\Sigma \text{L (total of all stream segments)}}{A \text{ (area of the basin)}} \tag{6}$$

4. Meandering ratio (M_r): It is ratio between straight and curved length of the primary stream in the catchment [2]. Therefore, higher M_r ratio reflects low run-off energy and higher sedimentation rate. M_r can be calculated as follows:

$$M_r = \frac{\text{L (straight)}}{L \text{ (curved)}} \tag{7}$$

5. Texture topography (T_t): This evidences the ability of a terrain to infiltrate water as it is controlled by rock types and structures in the catchment. It is calculated as the total number of streams (N_s) of all order in a basin per perimeter (B_p) of the basin [10]. Hence, T_t is calculated by the following equation [11]

$$T_t = \sum N_s / B_p \tag{8}$$

Smith [11] classified the texture topography as: very coarse (<2), coarse (2 to 4), moderate (4 to 6), fine (6 to 8) and very fine (>8).

3. Volumetric measures of the Lebanese rivers

It is significant to calculate the volume of water that enters the drainage system of rivers. This assists in characterizing the catchment ability to capture and outlet water. This requires elaborating quantitative analysis in the each catchment, where the volume of precipitated water and discharged water are measured.

For the precipitated water, data were collected from the available meteorological ground stations, as well as form remotely sensed products, with emphasis to Tropical Rainfall Mapping Mission - TRMM [12]; Climate Hazards group Infrared Precipitation with Stations - CHIRPS [13]; and from National Oceanographic Data Center – NOAA [14]. While, the discharge from the Lebanese rivers is periodically measured by the Litani River Authority (LRA) [15]. Therefore, water volume enters and outlet, along rivers, from each watershed was calculated as shown in **Table 3**.

Table 3 shows that the average volume water enters the catchments is about 480 million m^3/year, and the average discharge is approximately

No.	Catchment	Area (A)	Rainfall (R)	Discharge (D)	D/R	R/A
		km^2	Mm^3/year		%	Mm^3/km^2
1	El-Kabir R.*	303	260	222	—	0.38
2	Al-Bared R.	284	225	165	73	0.79
3	Abou-Ali R.	482	505	365	72	1.04
4	Ej-Jouz R.	196	125	80	64	0.64
5	Ibrahim R.	326	380	495	131	1.16
6	El-Kaleb R.	237	330	225	66	1.39
7	Beirut R.	216	260	100	38	1.20
8	Ed-Damour R.	333	335	255	76	1.00
9	El-Awali R.	291	320	280	88	1.09

No.	Catchment	Area (*A*)	Rainfall (*R*)	Discharge (*D*)	*D*/*R*	*R*/*A*
		km^2	Mm^3/year		%	Mm^3/km^2
10	Siniq R.	102	100	60	60	0.98
11	Ez-Zahrani R.	140	145	200	137	1.03
12	Litani R.	2110	2078	360	17	0.98
13	Al-Assi R.*	1930	1254	420	—	0.65
14	Hasbani-Wazzani R.*	645	598	225	—	0.89

**Transboundary River*

Table 3.
Volume of precipitated water discharged water from the Lebanese rivers' catchments [1].

247 million m^3/year. This indicates that the Lebanese rivers are discharging only about 51% of the precipitated where the rest 49% goes to the evapotranspiration and for groundwater recharge [1].

4. Water pollution in rivers

It is estimated that in Lebanon, more than 50% of water resources are under physiochemical and biological contamination, and rarely a source of water in Lebanon is pure [16]. Lately, the problem of water quality deterioration has become one of the major is striking challenges that acts on the national level, and it severely hurts human life. This includes mainly the pollution of surface and then followed by groundwater resources.

This unfavorable problem is being increased by the absence of the controls and therefore, disposal of liquid and solid wastes (i.e. industrial, municipal and agronomical wastes) is widespread, notably in river courses and streams. Hence, the bad geo-environmental situation, due to water pollution, added a challenging issue for the water sector in Lebanon, while it is surprising that no effective actions have been taken by the concerned governmental bodies and even there is unethical behavior by some inhabitants in different regions of the country.

The Litani River, the largest river in Lebanon, gives a typical example on water pollution in the country. This rivers, which includes more than 370.000 people in 246 towns and encompasses 174 km length, has been lately witnessing intensive pollution pressure. This implies the river Couse and its reservoir (i.e. Qaraaoun Reservoir). Therefore, it was described as "Death of a River" [17]. In this respect, pollution sources were determined including direct dumping of huge amounts of solid wastes and high volume of sewage water into the river course and its tributaries, excessive use of fertilizers and well as the presence of many landfills within the catchment of the river and thus acting on groundwater purity.

Many surveys and studies have been done to investigate water quality in Lebanon where some of them aimed also at identifying the sources of pollution. The largest number of these studies were either applied to selective regions, or sometime they were applied for a limited time period specific, while other studies investigated only one aspect of pollution (e.g. microbiological pollution).

It is still a paradox that even with the large number of studies, nothing has been improved yet in regard to water quality in all resources; besides the pollution level in being continuously increased. Moreover, no effective management plans have been addressed to resolve the problem, and if they are proposed/exist, they remain propositions or without creditable implementation.

There are many examples can be illustrated to expose the current situation on the deteriorated water quality in Lebanon. Below are some example [18]:

1. The analysis of selective water samples from the Litani River shows:

    - Nitrite (NO_2) 19 ppm (max. 0.1 ppm)
    - Chromium (Cr^{3+}) 0.27 (max. 0.05 ppm)
    - Staphylococcus 8750 (0 in 100 ml)
    - Total coliform 183,000 (0 in 100 ml)
    - Fecal coliform 180,000 (0 in 250 ml).

2. The analysis of water quality the Qaraaoun Reservoir shows [19–21]:

    Sodium (Na): 10 mg/l (WHO max. 200 mg/l)

    - Chromium (Cr) 0.02 mg/l (WHO max. 0.05 mg/l)
    - Zinc (Zn) 0.09 mg/l (WHO max. 3 mg/l)
    - Copper (Cu) 0.019 mg/l (WHO max. 2 mg/l)
    - Cylindrospermopsin toxin 1.7 μg/L (WHO guidelines 0.7 μg/l)
    - Cyanobacteria up to 200 μg/l (WHO limits 10 μg/l)
    - Carlson trophic state index 66 to 84 (CTSI max. 40).

3. The analyzed samples of groundwater analysis in different boreholes located in the Bekaa Plain showed that Nitrate (NO_3) concentration exceeded 300 mg L^{-1} [22].

4. The analyzed bottled water which were taken form 48 major water companies in Lebanon showed that approximately 80% were contaminated either chemically or biologically or combination of both [23].

Due to its significance and the resulted severe impact on human health and even life, there are some implementations done for waste disposal management, and thus several national and international projects were applied. Moreover, field campaigns, capacity building, inter-ministerial committees and business plans were established to identify the required measures and secure water quality and the existing ecosystems, but no enhancement in this concern could be touched yet [17].

5. Uncontrolled water pumping

In the view of shortened water supply besides an exacerbated demand, inhabitants are always searching for any available sources of water to compensate the difference between supply and demand. This primarily accounts the ease and the low-cost of exploitation of these sources. Hence, rivers are the most applicable resources to be invested in Lebanon, notably that these rivers and their major tributaries are widespread over short distances between urban clusters and among the arable lands where water is competitive. Hence, many rivers' tributaries are only few kilometers from each other.

In addition to this geographic aspect of the Lebanese rivers, there is no consolidated and effective environmental controls to regulate the behavior of people towards the exploitation of water from rivers. This can be also attributed to many other reasons including mainly the political situation in the country. Therefore, unfavorable works are widespread in all the Lebanese rivers, and illegal water abstraction from rivers plus dumping of liquid and solid wastes are commonly observed. These works can be summarized as follows:

1. Direct water pumping from rivers and the surrounding springs where this pumping does not follow and control or measuring approaches (i.e. illegal), and this is very common in upstream regions of the Lebanese rivers. The largest part (>90%) of the pumped water goes to irrigation (Example in **Figure 2**).

2. Uncontrolled water use from rivers where several private water systems are connected with rivers, and then conveying water along private-owned canals for irrigation. These canals can be for hundreds of meters long.

3. Chaotic water abstraction whether from the recharge zones of rivers, where rivers receive their water from, or from the recharge zones of springs and groundwater aquifers which in turn replenish rivers.

Figure 2.
Illegal pumping of water from rivers in Lebanon, a common observation.

6. Transboundary rivers

In spite of the small surface area of Lebanon (i.e. 10,452 km^2), the geography of the country makes its water resources shared with the neighboring regions. Therefore, about 2631 km^2 of Lebanon's surface area constitutes shared groundwater reservoirs with the neighboring regions, and this is equal to approximately 25% of the Lebanese area [2]. In addition, there are approximately 2878 km^2 (27.5%) of Lebanon's surface area comprises basins for transboundary rivers with the riparian regions. Thus, Lebanon contributes with major tributaries for three transboundary rivers [2]:

1. Al-Kabir River (222 million m^3/year from Lebanon): It almost represents the northern international boundary of Lebanon with Syria, and it encompasses a catchment area of 972 km^2 where 303 km^2 are in Lebanon.

2. Orontes River (420 million m^3/year from Lebanon): One of its major tributaries is originated from Lebanon and then named as Al Assi River. It is shared with Syria and Turkey with a total catchment area of 25,300 km^2 where 1930 km^2 are located in Lebanon.

3. Jordan Rivers (225 million m^3/year from Lebanon): In Lebanon the Hasbani-Wazzani River represents one of the primary tributaries of Jordan River which is shared with Syria, Jordan and Israel. The river, which is under frequent geo-political conflicts, has a total catchment area of about 18,425 km^2, only 645 km^2 of them are in Lebanon.

There are only 210 million m^3 of water which is used by Lebanon from shared water resources of the country [2], and this volume constitutes water pumped from transboundary rivers, as well as the estimated water abstracted from dug wells in shared groundwater aquifers). This volume represents only about 15% of the total volume of these resources.

In the lack to proper management of these shared water resources; however, Lebanon loses a significant portion of its water. According to Shaban and Hamzé [2], if Lebanon adopted integrated and appropriate management approaches for its shared water resources, and then work in the direction of utilizing around 50% of these resources; thus, a water volume of about 700 million m^3 can be added to the budget of water in Lebanon.

If this volume is allocated to consumers in the country; therefore, approximately 175 m^3 per capita (i.e. equivalent to about 80% of water demand per capita) will be added to water supply. Whereas, the benefit of about 60% of shared water in Lebanon; will totally provide the water demand per capita. This in turn catalysis the adoption of these resources which can be done by following integrated management strategies of these resources.

7. Dispute on dams construction

Lebanon is characterized by mountain topography where acute sloping terrain is dominant, and the majority of precipitation (rainfall and snow) accumulates on the elevated areas and then water rapidly flows along these slopes. The journey of water flow from mountains to the sea takes short time interval which was estimated approximately as few hours, notably that the average distance of flow does not exceed few tens of kilometers.

Therefore, the flow of water from the Lebanese mountains is considered as a major aspect of water loss and water outlet into the sea before any significant investment. Therefore, water harvesting should adopted as an alternative solution to tap water along the Lebanese rivers instead of water loss. Hence, construction of dams will be an optimal solution.

There are dams constructed in Lebanon along some rivers and major tributaries, but they are still of small-scale dams, except the ones of Qaraaoun and Shabrouh dams, which have capacity of 220 and 11 million m^3; respectively. In this regard, the

Ministry of Energy and Water (MoEW) obtained a long-term plan for surface water development within the horizon of 2030 where 18 dams are proposed.

It is unlikely that the Lebanese administration and public finance can accomplish the planned long-term plan for constructing the proposed dams and lakes before 2030. Reasons behind postpone/or obstruction of dams construction in Lebanon are tremendous. One of these reasons is the dispute on dams' construction, and more certainly the believe that the topography and geology of Lebanon are not suitable for dam's construction; in particular the seismic setting of Lebanon does not assure the stability of dams [1]. Therefore, there is always debate on dams' construction in Lebanon. Lately, a problem has been existed even on the political level where the two proposed dams of Bisri and Janeh regions have been come to the implementation phase. Thus, there is of postponing to start working on these dams due to the existed conflicts about their location suitability.

8. Conclusion

Rivers in Lebanon are compose the veins of agricultural development, notably that the largest portion of water supply is delivered from rivers, as well as rivers constitute about 42% of water resources in Lebanon besides 32% from groundwater and the rest 26% is from other sources and mainly springs. However, these estimates are still rough and the water cycle in Lebanon can be considered as a comprehensive cycle where all elements of water journey are included. Thus, rivers feed groundwater, and the later replenish spring and so on.

Rivers in Lebanon discharge about 2800 million m^3 per year, which a significant part of the water balance in Lebanon. Nevertheless, only 25% (or even less) of this amount is exploited and the rest is either lost to sea or shared with the neighboring regions.

The exploitation of water from these rivers is almost chaotic and illegal, notably for irrigation which consumes more than 70% of water in Lebanon. In addition to domestic and industrial uses, water from rivers is also used for hydro-power generation and it contributes to approximately 20% of electricity needs for the entire country.

There are several threats existed lately on these rivers and they include natural and man-made threats. Hence, the average annual discharge rate in rivers has been sharply decreased and some rivers showed decline in the discharge reach up to 60% of its normal discharge rate. This is attributed either to the direct pumping from these rivers or the over pumping form the feeding zone for springs and groundwater aquifers. In addition, the changing hydrologic regimes of the terrain surface plays a major role in controlling the amount of water in rivers [1].

Moreover, pollution is a major problem in all Lebanese rivers and rarely a river in Lebanon is found with pure water. Thus, outfalls from wastewater (e.g. sewages, domestic, etc.) sources and delivered towards rivers. This made some rivers, like the Litani River, as a source of contamination and diseases became widespread in the proximity of its tributaries. In addition, the physiochemical and bacteriological analysis of water and sediments in many of these rivers showed contamination exceeds several times the accepted standards.

The solution for the Lebanese rivers implies adopted an integrated management of river' water. This can be built in the context of a national water strategy where assessment and monitoring must be continuously applied, and this can be consolidated by creating environmental legislations and laws devoted for water in rivers of Lebanon.

Author details

Amin Shaban
National Council for Scientific Research, Beirut, Lebanon

*Address all correspondence to: geoamin@gmail.com

References

[1] Shaban, A. Water Resources of Lebanon. Springer Science Publisher. 2020. DOI: 10.1007/978-3-030-48717-1. 229 pp.

[2] Shaban, A., Hamzé, M. Shared water resources of Lebanon, Nova Publishing, NY. 2017. DOI:10.1007/978-3-030-48717-1. p 150.

[3] Schumm, S. The elevation of drainage systems and slopes in bad-lands at Perth Amboy, New Jersey. Geol. Soc. Amer. Bull., Vol. 67. 1956. pp. 597-646.

[4] Horton E. Drainage-basin characteristics. Trans. American Geophysical Union. 1932;**13**:350-361

[5] Pike, R., Wilson, S. Elevation-relief ratio. Hypsometric Integral and Geomorphic Area-Altitude Analysis." GSA Bull. 1971. 82: 1079-1084.

[6] Al Saud, Mashael.. Use of Remote Sensing and GIS to Analyze Drainage System in Flood Occurrence, Jeddah - Western Saudi Coast (Chapter in Book entitled: Drainage Systems. InTech Open Science. Croatia. 2012. 139-164pp. ISBN13: 978-953-51-0243-4.

[7] Raven, P., Holmes, N., Naura, M., Dawson, F. Using river habitat survey for environmental assessment and catchment planning in the U.K. Hydrobioloigia. 2000. Vol. 422(0):359–367.

[8] Morisawa, M. Geomorphology Laboratory Manual, John Wiley & Sons Inc. N.Y. 1976. pp.1-253.

[9] Shaban, A., Bou Kheir, R; Khawlie, M., Froidefond, J, Girard, M-C. Caractérisation des facteurs morphométriques des réseaux hydrographiques correspondant aux capacités d'infiltrations des roches au Liban occidental. Zeitschrift fur Geimorphologie. 2004. Vol. 48/1, pp 79-94.

[10] Horton R. Erosional developments of streams and their drainage basins: Hydro-physical approach to quantitative morphology. Geological Society of America Bulletin. 1845;**56**: 275-370

[11] Smith K. Standards for grading texture of erosional topography. American Journal of Science. 1950;**248**: 655-668

[12] TRMM (Tropical Rainfall Mapping Mission). Rainfall archives. NASA. 2015. <http:// disc2.nascom.nasa.gov/Giovanni/tovas/TRMM_V6.3B42.2.shtml>

[13] CHIRPS (Climate Hazards Group InfraRed Precipitation with Station data). 2017. Available at: http://chg.geog.ucsb.edu/data/chirps/

[14] NOAA (National Oceanographic Data Center). Lebanon Climatological Data. Library. 2013. Available at: http://docs.lib.noaa.gov/rescue/data_rescue_lebanon.html

[15] LRA (Litani River Authority). Rivers discharge records database.2017. Unpublished Report.

[16] Shaban, A. Physical and Anthropogenic Challenges of Water Resources in Lebanon. Journal of Scientific Research and Reports. 2014. Vol.3 Issue 3. 2014. 164-179.

[17] Shaban, A., Hamzé, M. The Litani River, Lebanon: An Assessment and Current challenges. Springer International Publisher. 2018. 179 p. DOI: 10.1007/978-3-319-76300-2.

[18] Nehme, N., Haidar, C. The Physical, and Chemical and Microbial Characteristics of Litani River Water.

Chapter in Book entitled: The Litani River, Lebanon: An Assessment and Current challenges. 2018. Springer. 179 p.

[19] Fadel A, Atoui A, Lemaire BJ, Vinçon-Leite B, Slim K. Environmental factors associated with phytoplankton succession in a Mediterranean reservoir with a highly fluctuating water level. Environ. Monit. Assess. 2015;**187**:633 https://doi.org/10.1007/s10661-015-4852-4

[20] Fadel, A., Faour, G., Slim, K., 2016. Assessment of The trophic state and Chlorophyll-a concentrations using Landsat OLI in Karaoun reservoir , Lebanon. Leban. Sci. Journal, 17, 130–145. 2016. https://doi.org/10.22453/LSJ-017.2.130145

[21] Fadel, A. , Slim, K. Evaluation of the Physicochemical and Environmental Status of Qaraaoun Reservoir. Chapter in Book entitled: The Litani River, Lebanon: An Assessment and Current challenges. 2018. Springer. 179 p.

[22] Darwich T, Atallah T, Baydoun S, Jomaa I, Kassem M. Environmental risk of nitrate accumulation in the soil-groundwater system in Central Bekaa Valley, Lebanon. 2014. International Conference on: The Water-Food-Energy-Climate Nexus in Global Drylands, Rabat, Morocco, 12-13 June 2014.

[23] CPA (Consumer Protection Association). Analysis of water quality in bottled water of Lebanon. 2018. Unpublished Report. 7 pp.

Chapter 7

Statistical Analysis of the Precipitation Isotope Data with Reference to the Indian Subcontinent

Supriyo Chakraborty, Siddharth Birmal, Pramit Kumar Deb Burman, Amey Datye, Fousiya A.A., Aravind G.H., PM. Mohan, Neha Trivedi and Rajendra K. Trivedi

Abstract

The isotopic analysis of precipitation provides useful information on a variety of hydrological and atmospheric processes. The dynamical characteristics of precipitation isotopes have been well investigated, but a systematic study of their statistical behavior seems to be lacking. We have performed the statistical analysis, basically the distribution characteristics of precipitation isotopes vis-a-vis rainfall data for specific regions. The probability distribution functions of precipitation isotopes have been calculated from local to global scales. It has been observed that the isotopic values, in general, followed a pattern that is similar to the normal distribution, though the rainfall distribution patterns are very different. Under certain circumstances, the isotopic distribution patterns closely resemble the normal distribution, implying a well constrained moisture source contributing to precipitation. The distribution patterns of oxygen and hydrogen isotopes on continental and global scales show similar behavior. It was observed that the distribution patterns of primary isotopic variables ($\delta^{18}O$ and δD) are not very sensitive to the outliers. On the contrary, the secondary parameter, d-excess, is very sensitive to outliers, which offers an effective means to quality control of the precipitation isotopic values.

Keywords: precipitation isotopes, Indian monsoon, probability distribution function, skewness, kurtosis

1. Introduction

Systematic collection and analysis of the isotopic content of precipitation across the globe have been carried out under the GNIP (Global Network of Isotopes in Precipitation) framework to study the temporal and spatial variabilities of the environmental stable isotopes. A large number of precipitation isotope data, mostly on a monthly timescale, is available with GNIP. Various investigators have used the data in understanding a variety of atmospheric and hydrological processes [1–8]. Specific statistical techniques have also been proposed to interpret the data [9, 10]. But one

IntechOpen

essential characteristic, viz. understanding the statistical behavior of the precipitation isotope data, did not receive much attention by the isotope hydrologists.

The probability distribution of atmospheric variables such as temperature, pressure, wind speed, precipitation, greenhouse gas concentrations, effective radiative forcing, aerosols etc. [IPCC 2013, [11]] is routinely used by the environmental scientists and operational meteorologists for different purposes. One of the primary applications is to assess the occurrence of extreme weather events or environmental conditions (see Chapter-1 of the IPCC Report [11]). Temperature variations usually exhibit a normal distribution pattern. Because of its symmetric behavior, it offers an intuitive method to assess the change in temperature characteristics for different periods or distinct spatial boundaries. For example, if the pattern shifts towards the right, say for a different time interval, then it would indicate an increase in mean temperature for that particular time interval (see Figure 1.8 in Ref: [11]). So the analysis of the distribution pattern of temperature over a long period provides an excellent means to assess the climate change.

Similarly, if the temperature distribution of a given region differs from the usual standard Gaussian pattern, then it may be concluded that the temperature in that particular region experiences different environmental controls. Unlike temperature, the *moisture* parameters such as rainfall, humidity, cloud size, cloud water content, etc. almost always deviate from normal distribution patterns [12]. Their distribution characteristics provide useful information on weather and environmental conditions and hence are extensively studied by the meteorologists.

Several investigators studied the statistical distributions of rainfall over the Indian region as well as from other parts of the world. The use of the probability distribution function of precipitation was described by Sharma and Singh [13]. Wither (2001). [14] and Nadarajah (2005). [15] examined the distribution characteristics of rainfall over sixteen locations in New Zealand and fourteen sites in Florida, respectively. They have also studied the maxima of rain throughout 1961–2001 for five areas in South Korea. Generalized extreme value distribution was fitted to those data to describe the anomalous distribution characteristic and, in turn, predict their future behavior. It has been shown that the annual maximum of daily rainfall in Japan closely followed the Weibull distribution pattern [16].

Ghosh et al. [17] studied the probability distribution of rainfall over Bangladesh. These authors observed that the generalized extreme value distribution provides a good fit to the rainfall data over Chittagong, Rajshahi, and Sylhet. However, the rainfall data from the Dhaka stations are best described in terms of Gamma distribution function [17].

In the Indian context, several people have reported the rainfall distribution characteristics. For example, Molay et al. [18] observed that the South West (S.W.) and North East (N.E.) monsoon rainfalls at individual representative stations in India followed the Gamma distribution pattern. Statistical analysis of rain for Uttarakhand, India, was carried out by Vikram and Jahangeer [19]. These authors observed that the rainfall data covering 46% of all the districts is best described by the Weibull distribution function, followed by the Chi-squared and log-Pearson distribution patterns. Statistical properties of rainfall at Coimbatore in South India were discussed by Lavanya et al. [20].

The isotopic composition of precipitation, an essential attribute of the rainfall variability, is widely studied by the hydro-meteorologists for understanding a variety of processes. In the Indian context, one of the first comprehensive analysis was carried out based on the precipitation isotope data across the country by Bhishmkumar et al. [21]. The local meteoric water lines were calculated for several locations. They show significant variations in slope and intercept across the continent. Several other works were undertaken to address a variety of issues related to hydrological and atmospheric processes [22–26]. One of the most intriguing aspects is the relation between rainfall amount and its isotopic values; the so-called *amount effect* has been examined for different regions and for varying time scales. Isotopic values are essentially derived from rainwater, so their physical properties are expected to mimic that of the rainwater. However, deviations were observed mostly in their dynamical behavior. For example, during the summer monsoon season (June to September) in India, the wind is primarily south-westerly. After the monsoon withdrawal in early October, the wind is pre-dominantly north-easterly. Because of this reversal in the wind and hence in the moisture source, the isotopic values of precipitation in southern peninsular India experience strong seasonality [27]. But this characteristic is not well manifested in the rainfall pattern. One pertinent question may be posed: do the statistical distributions of precipitation isotopes also follow a similar pattern of the rainfall distribution? If it is not, what are the implications of having a different kind of distribution pattern? Do the distribution patterns show significant spatial variability? Since the isotopic values of proxy records are widely used for paleoclimatic investigation, especially the rainfall reconstruction, do the statistical distribution patterns of precipitation isotope have any implication in this regard? In this work, we characterize the probability distributions of precipitation isotopic values and endeavor to address the above-mentioned issues.

2. Data and methodology

Precipitation isotopic data on a daily scale has been collected from several locations in India. These sites are Port Blair in the Andaman Islands (data available: 2012–2018; [26, 28–30]), Minicoy Island in the Arabian Sea (2015–2018; [31]); Kolkata in east India (2015–2018; [26]); Darjeeling (2013–2018) and Tezpur in northeast India (2016–2018; [26]); Ahmedabad (2007) [25] and Pune in west India (2014–2018; [32, 33]), Bandipora in north India (July-2016 to June-2018; [34]) and Kozhikode in south India [22]. Precipitation isotope data of a few sites in Bangladesh have also been used [29, 35]. **Figure 1** shows the precipitation sampling sites and the core monsoon zone of India (shaded). Additionally, precipitation isotope data across the tropics have also been used to examine the statistical behavior of the isotopic data on a global scale. The data is available in [29].

Rain-gauge data have also been used to study the distribution pattern of rainfall. To have regional-scale rainfall variability, the TRMM (Tropical Rainfall Measuring Mission) satellite-derived gridded rainfall data available at 0.25° × 0.25° resolution [36] was used. We have also used the oxygen isotopic values of the speleothem that had been used to investigate the monsoon variability on the past timescale. Berkelhammer et al. [37] reported the oxygen isotopic time series of two speleothem samples belonging to the core monsoon zone (CMZ) of India. A composite $\delta^{18}O_{speleothem}$ record was prepared by them spanning from AD 625 to A.D. 2007. The CMZ consists of a wide area (approx. 18-27°N, 69-88°E; see **Figure 1**) in central India, and the rainfall in this region maintains a strong correlation with the all India summer monsoon rainfall [38]. Hence the speleothem derived rain is believed to be representing the large scale rainfall pattern over India.

The isotopic ratio is expressed in δ notation and is defined as the relative difference between the heavy to light isotopic values (i.e., $^{18}O/^{16}O$) of the sample and reference material. The difference is normalized with the isotopic ratio of the reference material and then multiplied by a factor of 1000. The isotopic value is denoted in δ notation and expressed in permil (‰). As an example, the oxygen isotopic value of precipitation is defined as follow: $\delta^{18}O\ (‰) = [(^{18}O/^{16}O)_{precipitation} - (^{18}O/^{16}O)_{reference}]/[(^{18}O/^{16}O)_{reference}] * 1000$.

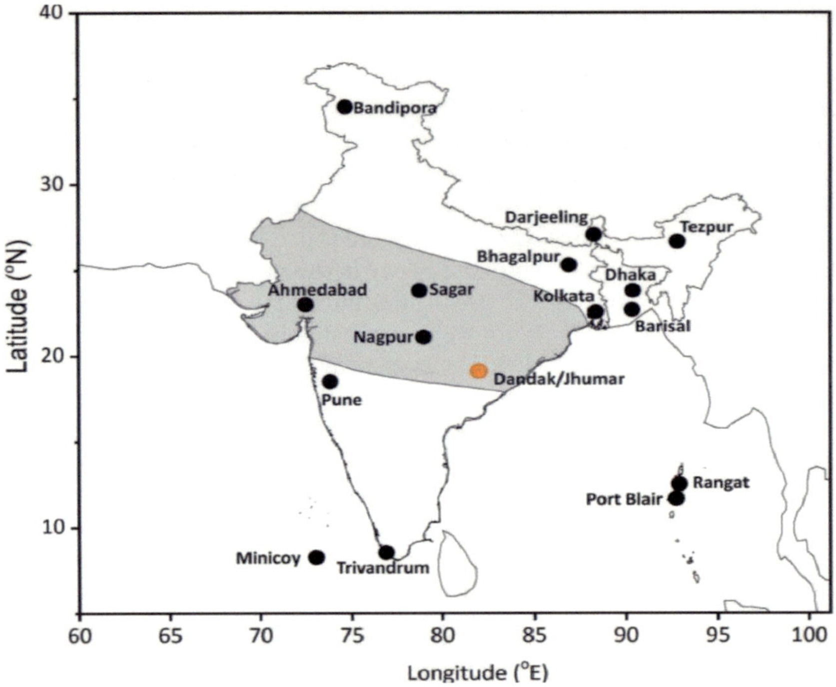

Figure 1.
Map showing the rainwater sampling sites across India. Both published and unpublished data were used; for details, see section 2. The shaded area shows the core monsoon zone of India. The orange dot indicates the speleothem sample location reported in Berkelhammer et al. [37].

The reference material is the Vienna Standard Mean Ocean Water (VSMOW), whose value has been assigned to 0‰ (IAEA 2009) [39]. In tropical regions, precipitation $\delta^{18}O$ typically varies from 0 to −15‰.

A secondary parameter, d-excess [40] is commonly used to identify the source moisture. It is defined as follows: d-excess = $\delta D - 8 * \delta^{18}O$.

It is to be noted that the control mechanisms on $\delta^{18}O$ or δD, and d-excess are not necessarily the same. Isotopic ratios ($^{18}O/^{16}O$ and D/H) are controlled by the mass-dependent fractionation process. They typically show a strong correlation indicating that they are controlled by similar environmental processes [41]. But, since d-excess is a difference between these two isotopic values, the individual fractionation effect is likely to be minimized in normal environmental conditions. The d-excess is mainly controlled by kinetic processes, say during evaporation of water from the seas, in which the environmental condition, such as relative humidity, plays a significant role. On the other hand, the evaporation of raindrops causes enrichment in heavier isotopes. But since the rates of escape of D and ^{18}O differ, which is a function of their distribution coefficient, the d-excess would decrease. So intense evaporation would cause a significant change in d-excess values. Because of these physical processes, the distribution patterns of the isotopic values and d-excess may be different.

2.1 Skewness and kurtosis

Skewness (S) is a statistical parameter that is used to examine the symmetry of a distribution. A distribution is said to be symmetrical if the frequencies are symmetrically distributed about the mean [42]. For a normal distribution, the S value is zero.

Kurtosis (K) is defined as the fourth standardized moment. It is a measure of the relative flatness of the distribution function. For a normal distribution K = 3, based on which *excess kurtosis* is used in which three is subtracted from the observed kurtosis [42]. If the excess kurtosis is negative for a given distribution, then it implies that a particular distribution contains less number of extreme values compared to a dataset that is normally distributed. Analysis of kurtosis provides a useful means to identify extreme values. We will discuss how this parameter could be used to differentiate the extreme values (occurring naturally) from the outliers (the probable reason for an analytical artifact). To calculate S and K, both the rainfall and the isotopic data of all the years are separately clubbed together to form two combined time series. Separate analyses were done for local scale (Port Blair and Pune), regional scale (northeast India and Bangladesh), continental-scale (entire India and Bangladesh), and global scale (the whole tropics). The S and K values were calculated using the OriginPro software, where bin size is decided by default; Excel-based SKEW() and KURT() function were also used for this purpose.

The distribution pattern of the isotopic profiles is examined on a wide scale. Isotopic data from Port Blair were used to study the pattern on a local scale. Similarly, isotopic data from Pune and its surrounding region, representing a relatively larger spatial scale, were used for studying the distribution pattern. Precipitation isotopic values of a vast area in northeast India, including Bangladesh, were used to study the distribution pattern on a regional scale. Finally, a large dataset spanning approximately from A.D. 2013 to A.D. 2017 from around the tropics was used to examine the isotopic distribution pattern on a global scale.

3. Results and discussion

3.1 Distribution characteristics of precipitation isotopes

3.1.1 Site: Port Blair

Figure 2(a) and **(b)** show the probability distribution functions (PDF) of rainfall and its oxygen isotopic values, respectively. At the same time, **Figure 2(c)** and **(d)** show the PDFs of the hydrogen isotopic ratio (δD) and d-excess, respectively. The values of S, K, and the number of samples (N) for each distribution are shown in the respective panels. The normal distribution curve in each case, shown as a thin line, has been drawn using the mean and standard deviation of the respective distribution.

The Bay of Bengal region experiences cyclonic disturbances typically during the pre-monsoon and post-monsoon seasons. Heavy to very heavy rainfalls are received during these events, mostly during the post-monsoon season, which is associated with highly depleted oxygen isotopic values, occasionally registering as low as −17 to −18‰ [28]. Hence, to avoid these extreme rainfall events and, in turn, the highly depleted isotopic values, isotopic distribution patterns for the summer monsoon season (May to September) have been redrawn. **Figure 3** depicts the corresponding rainfall (a), $\delta^{18}O$ (b), δD (c) and d-excess (d) distribution patterns.

The annual scale Port Blair $\delta^{18}O_{precipitation}$ shows a negatively skewed distribution pattern (**Figure 2(b)**). $\delta^{18}O_{precipitation}$ typically ranges from −14 to 2‰. A few high positive values up to +4‰ are observed, which mostly occur during the relatively dry environment prior to monsoon onset in May. But towards the negative side, low values up to −18‰ are observed, which make the distribution pattern left asymmetric, yielding a skewness of −1.16, and kurtosis: 2.78. Incidentally, the rainfall distribution pattern is highly positively skewed, with S and K values being 3.04 and 12.63, respectively (**Figure 2(a)**).

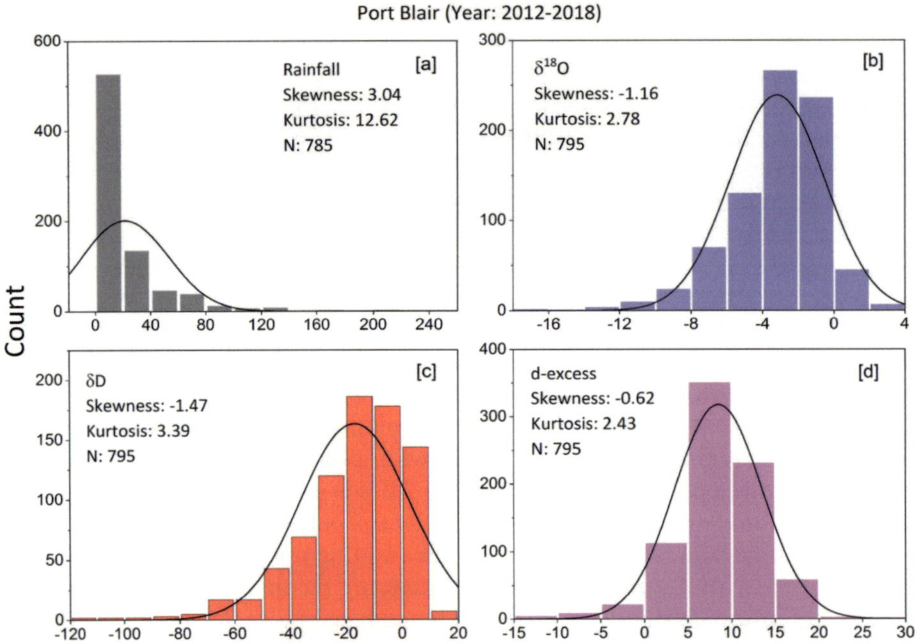

Figure 2.
Distribution characteristics of rainfall and their isotopic values at Port Blair for the time period of 2012–2018. (a): Rainfall; (b): $\delta^{18}O$; (c): δD; (d): d-excess. Y-axis represents the count. The thin gray line is the normal distribution curve drawn based on the respective mean and standard deviation values.

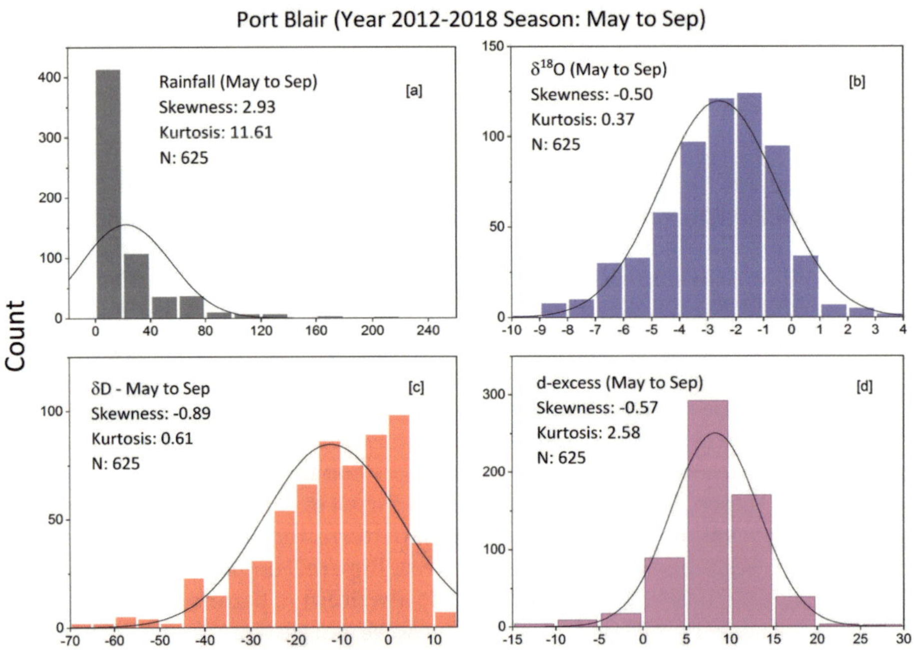

Figure 3.
Same as ***Figure 2*** *but for the summer monsoon season (May to September).*

A high value of kurtosis means the occurrence of intermediate values (falling within ±1 sigma) is less likely, but the extreme values are more likely. Hence, in this case, the high positive kurtosis indicates a significant amount of high to very high rainfall (≥ 100 mm/day) events populating in the tail region of the distribution. According to the amount effect concept, high rainfall produces more depleted isotopic values. Hence the distribution patterns are expected to show opposite behavior, somewhat mirror image to each other.

We have also examined the distribution patterns during the monsoon season (May to Sep) in order to eliminate the effect of high rainfall events arising due to intense cyclonic activities. **Figure 3** shows the distributions. In the case of rainfall distribution (**Figure 3(a)**), the K value decreases marginally (12.63 to 11.61). Still, the same for $\delta^{18}O_{precipitation}$ decreases significantly from 2.78 to 0.37, implying the effect of the heavy rainfall events on the isotopic value is eliminated. δD also shows a similar behavior, though the K values of d-excess in both cases are similar (2.43 and 2.58, respectively). The *residual kurtosis* in the d-excess cases differs only marginally, i.e., by −0.57 and −0.42, respectively, from the normal distribution. These imply the d-excess distribution pattern to have a close resemblance with a normal distribution and be less sensitive to the intense rainfall events, unlike $\delta^{18}O$ and δD.

3.1.2 Site: Pune

Pune belongs to a semi-arid region in western India. Being in the leeward side of the Western Ghats mountain range, the area is characterized by rain shadow region. As a result, the raindrop evaporation is significant, which makes the isotopic values relatively enriched in heavier isotopes. The distribution patterns of the above mentioned four parameters are shown in **Figure 4**. The S and K values of precipitation for the Pune region also show high values, 3.38 and 13.69, respectively (**Figure 4(a)**) comparable to that of the Port Blair values, implying a positively asymmetric distribution pattern. δD and $\delta^{18}O$ show moderate values of skewness and kurtosis (**Figure 4(b)** and **(c)**, respectively).

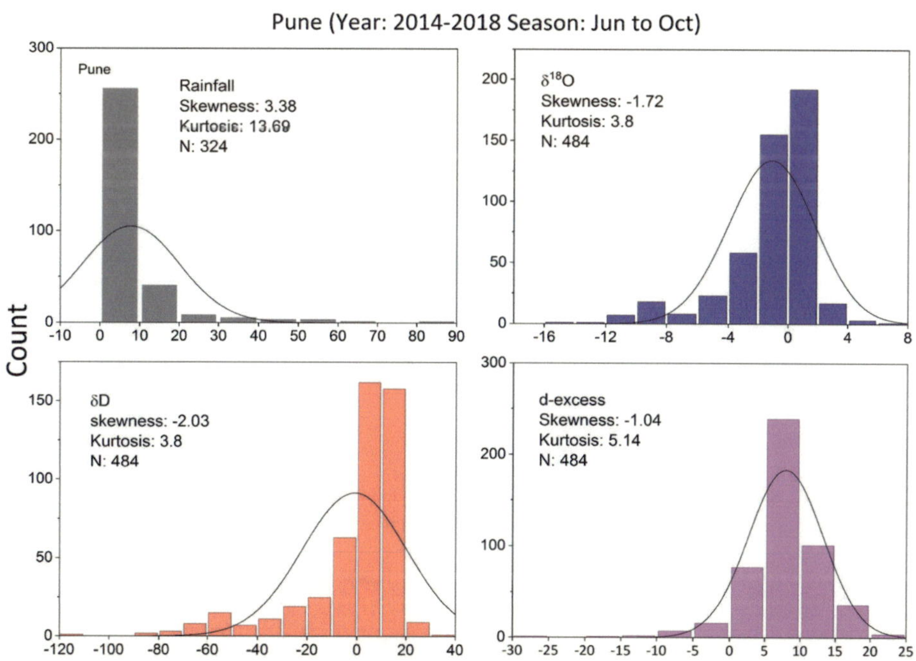

Figure 4.
*Same as **Figure 3** but for the Pune region.*

3.1.3 Northeast India and Bangladesh

This region contains a large forest area and receives one of the highest rainfall amounts in the world. As a result, the climate is humid subtropical. December to January is dry season, and little rain is received during this time. However, a significant amount of rainfall is received during the pre-monsoon season (Mar to May) due to the thunderstorm activities. The monsoon onset takes place in early June and continues until early October. October to December is considered as the post-monsoon season. The precipitation isotope analysis for this region has been carried out by several investigators. We have used those data available since A.D. 2007. Additionally, we have analyzed samples from two sites, Darjeeling and Tezpur. The records are available from A.D. 2013 to A.D. 2018 for Darjeeling and A.D. 2015 to A.D. 2018 for Tezpur. All the data have been compiled, and frequency distribution plots of $\delta^{18}O$ and d-excess were generated. Since δD has a distribution pattern very similar to the pattern of $\delta^{18}O$, their distribution patterns are not shown. Similarly, the rainfall distribution patterns are similar to other sites, and hence those plots are not shown here. **Figure 5** shows the season-wise distribution patterns of $\delta^{18}O$. The plot for d-excess has been shown only for the pre-monsoon season.

The distribution patterns of $\delta^{18}O_{precipitation}$ for this region have been shown (**Figure 5**) for three seasons. **Figure 5a** shows the $\delta^{18}O_{precipitation}$ pattern for the pre-monsoon (Mar to May) season, including Jan and Feb. The low value of S (0.48) yields a near-normal distribution of $\delta^{18}O_{precipitation}$. And a very low K value (0.002) means a near absence of extreme values. Similarly, d-excess also shows nearly a symmetric distribution (**Figure 5(b)**), and a low K value implies the frequency of extreme d-excess is negligible. Interestingly, other seasons, that is, monsoon (**Figure 5(c)**) and post-monsoon (**Figure 5(c)**) season, do not show such behavior in the distribution pattern. In both cases, the distribution patterns are negatively skewed. The reason for such kind of behavior may be explained as follow.

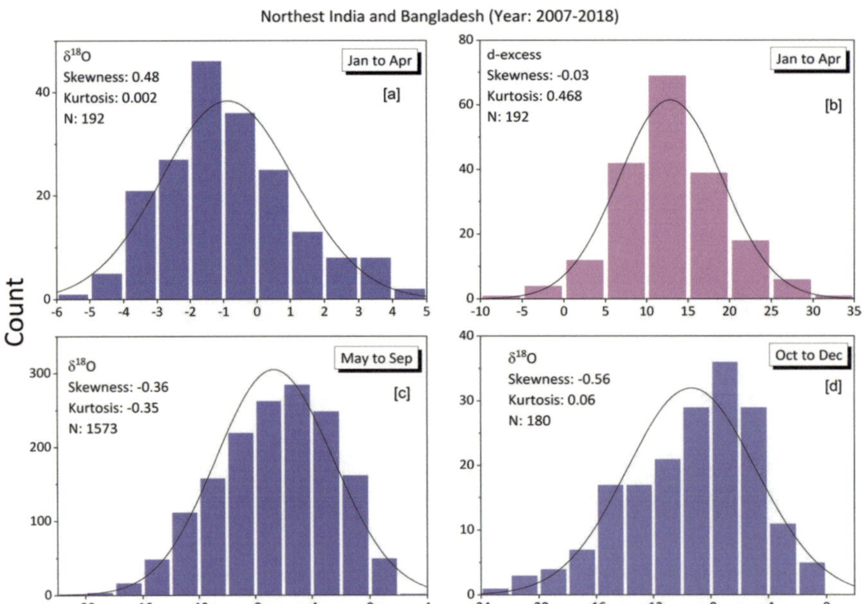

Figure 5.
Oxygen isotopic distribution pattern for the Northeast India and Bangladesh regions. (a) January to April, (b) same for d-excess, (c) same for $\delta^{18}O$ but for May to September season, (d) same for $\delta^{18}O$ but for the post-monsoon season (Oct to Dec).

Northeast India receives a considerable amount of rainfall during the pre-monsoon time, mainly due to the thunderstorm activities. Choudhury et al. [43] demonstrated that a large amount of recycled water vapor, mainly generated by the evapotranspiration processes, contributes to rainfall during this time. The contribution of marine vapor during this time is negligibly small. Hence the moisture source is reasonably well constrained, implying that the isotopic variability of precipitation would be limited. Mean and standard deviation are −0.86 and 1.99‰, respectively. The extreme values −6 and + 5‰ are nearly equidistant from the mean value. Hence, a near-normal distribution of $\delta^{18}O_{precipitation}$ results in this season.

On the other hand, the d-excess for this season also shows a very similar behavior. It possesses low S (−0.03), and K (0.047) values, and the mean value (12.86‰) is equidistant from the extreme values of −10 and + 35‰; hence a near-perfect normal distribution plot is produced. During the advent of the monsoon season in early June, a huge amount of marine vapor sweeps across this region. With the progress of the monsoon, the moisture source is also believed to be shifted towards the South Bay of Bengal [44]. These long-range transports cause further depletion in precipitation isotopic values. On the other hand, though the extent of recycled vapor is reduced considerably, it still contributes to some extent. Because of these multiple vapor sources, the isotopic values of precipitation suffer a wide variability.

Additionally, intense convective activities during the monsoon season cause large depletions in isotopic values. As a result, the precipitation isotopes may have values lower than −20‰ (**Figure 5(c)**). This characteristic feature drives the isotopic distribution pattern and makes it asymmetric. It is also known that intense convective activities cause high d-excess [28] due to the feeding of recycled moisture into the cloud system [45]. The northeast region experiences such kind of convective activities quite often during the monsoon season but not much during the pre-monsoon season [46]. This process is also likely to contribute to yielding a relatively high K value for the monsoon season. The post-monsoon season, on the other hand, gets moisture from multiple sources [35], which also makes the corresponding distribution pattern skewed in the negative direction.

3.2 Continental and global scales

Precipitation isotopic distribution patterns ($\delta^{18}O$ and d-excess) on a continental scale covering India and Bangladesh are presented in **Figure 6(a,b)**. The lower panel plots **Figure 6(c,d)** show the same approximately for the entire tropical region.

In this context, it is to be noted that the precipitation isotope data were obtained from nineteen sites across the tropics, including an extra-tropical area Nagoya, in Japan. Description of the sites and data are available in [29]. When the K value of d-excess was calculated for the entire dataset, an unusually high value (117) was obtained. A careful analysis of individual sites revealed that one particular location, Mulu, in Malaysia, was mainly responsible for producing such a high value. Unrealistically high values of d-excess (>50) were detected in several instances. These values were not associated with extreme rainfall events; average rainfall during these events was 24.9 mm/day. A running kurtosis analysis was carried out on the d-excess dataset. A total of 16 events (out of 1091 data points) were found, which showed sharp changes in K value. When these anomalous data points were excluded, the K value dropped to 1.44. So the anomalous values may have occurred due to analytical error. Because of this reason, the Mulu site data set were excluded. Similarly, two events in Kuala Terengganu, Malaysia, and one event at Windhoek, Namibia, showed anomalous K values. Hence they were also excluded. Recalculation yielded a K value of 3.58. In this context, it may be noted that extreme rainfall events do not necessarily create extreme value in d-excess. This is apparent

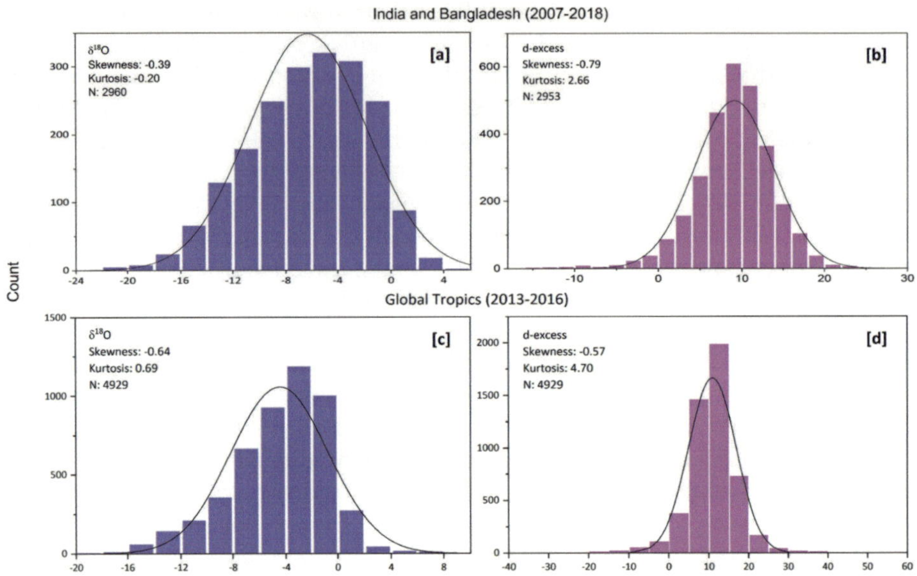

Figure 6.
Distribution pattern of precipitation isotope on the continental scale. (a) and (b) show the $\delta^{18}O$ and d-excess for India and Bangladesh. The same for the global tropics has been demonstrated in (c) and (d), respectively.

in the case of Port Blair data. When the Oct-Dec data series were excluded as this season is prone to intense cyclonic disturbances, the K value changed only from 2.78 to 2.58. Hence, the analysis of kurtosis may be used in separating the extreme events from the outliers caused by the experimental artifact. However, more study is required to ascertain this characteristic.

The continental-scale distribution pattern of $\delta^{18}O_{precip}$ (India and Bangladesh) and that of the global tropics show similar characteristics. In both cases, the $\delta^{18}O_{precipitation}$ shows a similar range, approx +8 to −20‰. But skewness and kurtosis in the former case (−1.06, 0.53) differ slightly relative to the global case (−0.64, 0.69). On the other hand, the d-excess ranges approximately from −20 to +30‰ in both cases. Low S values illustrate symmetrical distribution patterns. But the K values differ to some extent. The residual kurtosis in the case of India and Bangladesh (−0.34) is lower than that for the global tropics (+0.58). This means the d-excess distribution for the global case is relatively more populated by extreme values. Since d-excess primarily depends on the moisture source, a high proportion of extreme values means rapid changes in moisture sources, or in other words, rapid changes in circulation patterns. This is likely for the tropical region, as the tropics basically serve as the meeting point of the inter-hemispheric circulations, which is the inter-tropical convergence zone. On a shorter spatial scale, such as the Indian subcontinent, the effect is expected to be moderate.

3.3 Paleoclimatic implication

Here we discuss if there is any implication in analyzing the statistical behavior of the precipitation isotope variation. Isotopic analysis of certain natural archives is widely used to reconstruct the past monsoon rainfall variability. Speleothems or cave deposits have been extensively used for this purpose [37, 47–50]. The basic premise of this approach is that the isotopic signal of the ancient rainwater is preserved in the oxygen isotopic composition of calcium carbonate of the speleothem samples. But, the rainwater that percolates through the bedrock may have a residence time of a few years, typically 7–8 years. Hence the rainfall signal that is obtained from speleothem isotopic analysis represents an integrated effect of a few years. So the large variability of precipitation isotope would be attenuated, and a mean signal would be obtained. Hence, the variability of $\delta^{18}O_{speleothem}$ is expected to be much less than that of the $\delta^{18}O_{precipitation}$. Similarly, the extreme values in $\delta^{18}O_{speleothem}$ are also likely to be low because of the spatial and temporal integrations of the precipitation isotopic values. As a result, the $\delta^{18}O_{speleothem}$ would possess low skewness and kurtosis values to yield a near-normal distribution. We have examined this hypothesis using a speleothem isotopic record. Berkelhammer et al. [37] produced a 1400 years record of $\delta^{18}O_{speleothem}$ belonging to the core monsoon zone of India and demonstrated that this record was able to make monsoon variability for the last 1400 years. The distribution pattern of $\delta^{18}O_{speleothem}$ is shown in **Figure 7(a)**. A near-zero skewness (0.025) means the distribution pattern is highly symmetric.

Similarly, a low kurtosis value indicates that the central and the intermediate values mainly populate the distribution, while the extreme values play a minor role. In other words, the mean rainfall pattern is well captured. Now, a question may be raised: what would be an ideal precipitation isotope distribution pattern of the CMZ that would reliably be recorded by the speleothem samples? Obviously, if the anticipated pattern possesses a near-normal distribution, the signal would be expected to be accurately recorded by the speleothem sample. But, if the same pattern deviates widely, meaning high values of skewness and kurtosis characterize it, then the signal recorded in the speleothem sample could be noisy. So to get a realistic rainfall reconstruction from a proxy sample, the isotopic values of the rainfall in the neighboring region are expected to have near-normal distribution patterns. We have collected precipitation samples at Sagar and Nagpur approximately for the time period of 2016 to 2019 and used precipitation isotope data of Ahmedabad [25], all of whom belong to the CMZ of India (**Figure 1**). The $\delta^{18}O_{precipitation}$ distribution pattern of these samples is shown in **Figure 7(b)**. The low values of skewness and kurtosis imply that the pattern is indeed near normal. Hence the choice of speleothem sample from the CMZ seems to serve well as far as the rainfall reconstruction is concerned.

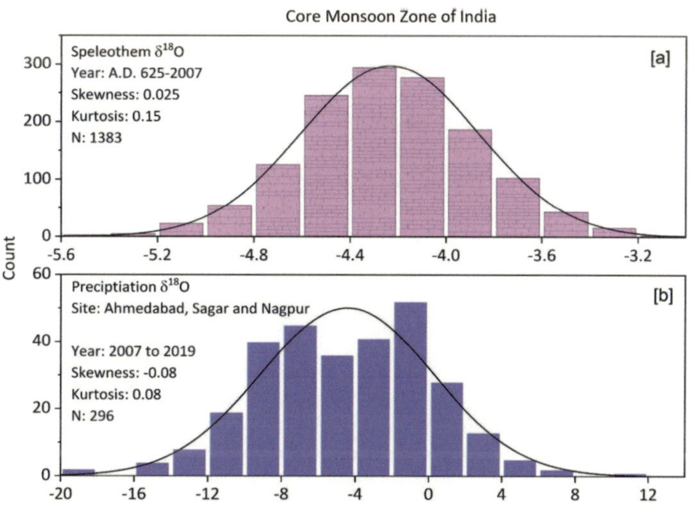

Figure 7.
Distribution patterns of (a) speleothem $\delta^{18}O$ and (b) $\delta^{18}O$ precipitation belonging to the core monsoon zone (CMZ) of India.

4. Summary and conclusions

Statistical analysis of precipitation isotopes was carried out on a large number of samples with a variety of spatial scales. Precipitation samples almost always showed a positively skewed pattern with high values of skewness and kurtosis, typically 2 to 3 and 11 to 13, respectively. On the other hand, the oxygen isotopic values of precipitation mostly showed small negative skewness (typically −0.5 to −1.0). The kurtosis value varied from 0 to 4. The secondary parameter d-excess is characterized by low negative skewness (−1 to 0), and its kurtosis ranges from 0.5 to 5. A comparison of these statistical parameters among precipitation and its isotopic values reveals that precipitation pattern is always skewed and contains a significant amount of extreme values. But the isotopic distribution patterns are near-symmetrical, and they are not very sensitive to extreme rainfall events. This may mean that the isotopic values are less constrained by environmental forcing in comparison to the precipitation. Hence the mean values of precipitation isotopes may be estimated/predicted with better precision than the precipitation. This may have application in paleoclimatic reconstruction.

It was observed that the kurtosis value of d-excess is very sensitive to outliers that may have occurred due to analytical error, but similar behavior was not found in the case of $\delta^{18}O$ or δD. Hence, the distribution parameters may offer a means to better quality control of the precipitation isotope data.

One of the potential applications of this technique is to study the interannual variability of monsoon rainfall. If a significant change is found especially in K value of d-excess distribution pattern in the year to year rainfall, then that change could imply different dynamical control and hence the monsoon circulation.

Acknowledgements

IITM is fully supported by the Earth System Science Organization of the Ministry of Earth Science, The Govt. of India. We thank the Director, IITM and R. Krishnan, Executive Director, Centre for Climate Change Research, IITM, for their support and encouragement.

Conflict of interest

The authors declare no conflict of interest.

Author details

Supriyo Chakraborty[1,2]*, Siddharth Birmal[2], Pramit Kumar Deb Burman[1,2], Amey Datye[1], Fousiya A.A.[3], Aravind G.H.[4], PM. Mohan[5], Neha Trivedi[6] and Rajendra K. Trivedi[6]

1 Indian Institute of Tropical Meteorology, Ministry of Earth Science, Pune, India

2 Department of Atmospheric and Space Science, Savitribai Phule Pune University, Pune, India

3 Department of Geology, Anna University, Chennai, India

4 Geology Department, Central University of Kerala, Periye, India

5 Pondicherry University, Port Blair, India

6 Department of Applied Geology, Dr. Harisingh Gour Vishwavidyalaya, Sagar, India

*Address all correspondence to: supriyo@tropmet.res.in

References

[1] Araguás-Araguás, L., Froehlich, K., and Rozanski, K., Stable isotope composition of precipitation over southeast Asia, J. Geophys. Res., 103(D22), 28721-28742, doi:10.1029/98JD02582. (1998)

[2] Feng, X., Faiia, A. M., and Posmentier, E. S. Seasonality of isotopes in precipitation: A global perspective, J. Geophys. Res., 114, D08116, doi:10.1029/2008JD011279.(2009)

[3] Liu, J., Fu, G., Song, X., Charles, S. P., Zhang, Y., Han, D., and Wang, S. Stable isotopic compositions in Australian precipitation, J. Geophys. Res., 115, D23307, doi:10.1029/2010JD014403.(2010)

[4] Aggarwal, Pradeep K.; Romatschke, Ulrike; Araguas-Araguas, Luis; Belachew, Dagnachew; Longstaffe, Fred J.; Berg, Peter; Schumacher, Courtney; and Funk, Aaron, "Proportions of Convective and Stratiform Precipitation Revealed in Water Isotope Ratios"Earth Sciences Publications. 13. https://ir.lib.uwo.ca/earthpub/13/ (2016)

[5] Rahul, P., P. Ghosh, S. K. Bhattacharya, and K. Yoshimura (2016), Controlling factors of rainwater and water vapour isotopes at Bangalore, India: Constraints from observations in 2013 Indian monsoon, J. Geophys. Res. Atmos., 121, 13, 936-13, 952, doi:10.1002/2016JD025352.

[6] Lekshmy, P.R., Midhun, M., Ramesh, R. Influence of stratiform clouds on δD and δ18O of monsoon water vapour and rain at two tropical coastal stations, J. Hydrol. 563, 354-362, doi: 10.1016/j.jhydrol.2018.06.001 (2018).

[7] Jeelani, G., Deshpande, R., Galkowski, M., Rozansky, K. Isotopic composition of daily precipitation along the southern foothills of the Himalayas: impact of marine and continental sources of atmospheric moisture, Atmos. Chem. Phys. 18, 8789-8805 (2018)

[8] Ahmed, N., Kurita, N., Chowdhury, M. et al. Atmospheric factors controlling stable isotope variations in modern precipitation of the tropical region of Bangladesh, Isotopes in Environmental and Health Studies 56, 220-237 (2020).

[9] Crawford Jagoda, Hughes Catherine E., Lykoudis Spyros. Alternative least squares methods for determining the meteoric water line, demonstrated using GNIP data,Journal of Hydrology,Volume 519, Part B,doi.org/10.1016/j.jhydrol.2014.10.033 (2014)

[10] Nguyen Le Duy, Ingo Heidbüchel, Hanno Meyer, Bruno Merz and Heiko Apel What controls the stable isotope composition of precipitation in the Mekong Delta? A model-based statistical approach. Hydrol. Earth Syst. Sci., 22, 1239-1262,doi.org/10.5194/hess-22-1239-2018 (2018)

[11] IPCC, Climate Change (2013) The physical science basis. In: Stocker TF, Qin D, Plattner G-K, Tignor M, Allen SK, Boschung J, Nauels A, Xia Y, Bex V, Midgley PM (eds) Contribution of working group I to the fifth assessment report of the Intergovernmental Panel on Climate Change. Cambridge University Press, Cambridge, New York, 1535 pp

[12] Stull, R. Meteorology for Scientists and Engineers. Brooks/Cole, Cengage Learning, 501p. (2000)

[13] Sharma, M., & Singh, J. Use of Probability Distribution in Rainfall Analysis. New York Science Journal 2010;3(9) (2010)

[14] Withers, Christopher S., and Saralees Nadarajah. Evidence of Trend

in Return Levels for Daily Rainfall in New Zealand. Journal of Hydrology (New Zealand), vol. 39, no. 2, 2000, pp. 155â "166., https://www.jstor.org/stable/43944839?seq=1

[15] Nadarajah, S. Extremes of Daily Rainfall in West Central Florida. Climatic Change 69, 325-342 https://doi.org/10.1007/s10584-005-1812-y (2005)

[16] Hirose, H. "Parameter estimation in the extreme-value distributions using the continuation method." Transactions of Information Processing Society of Japan, 35(9). (1994)

[17] Ghosh S., Kumar Roy Manindra, Chowdhury Biswas Soma Determination of the Best Fit Probability Distribution for Monthly Rainfall Data in Bangladesh, American Journal of Mathematics and Statistics,6(4): 170-174 DOI: 10.5923/j.ajms.20160604.05 (2016)

[18] Mooley D. A. and Rao G. A. Distribution Function for Seasonal and Annual Rainfall over India. Mon. Wea. Rev., 99, 796-799, https://doi.org/10.1175/1520-0493(1971)099<0796:DFFSAA>2.3.CO;2. (1971)

[19] Kumar Vikram; Shanu Jahangeer Statistical distribution of rainfall in Uttarakhand, India Applied Water Science, Volume 7, Issue 8, pp.4765-4776, DOI:10.1007/s13201-017-0586-5,(2017)

[20] S. Lavanya, M. Radha and U. Arulanandu, Statistical Distribution of Seasonal Rainfall Data for Rainfall Pattern in,TNAU1 Station Coimbatore, Tamil Nadu, India, ,Int.J.Curr.Microbiol.App.Sci,7(4): 3053-3062. https://doi.org/10.20546/ijcmas.2018.704.346 (2018)

[21] Kumar Bhishm, Rai S.P., Saravana Kumar U., Verma S.K., Garg P., Vijay Kumar S.V., Jaiswal R., Purendra B.K., Kumar S.R., Pande N.G.. Isotopic characteristics of Indian precipitation. Water Resources Research. 46, W12548, doi:10.1029/2009WR008532. (2010)

[22] Datta P.S., Tyagi S.K., Chandrasekharan H. Factors controlling stable isotope composition of rainfall in New Delhi, India,Journal of Hydrology, Volume 128, Issues 1-4,Pages 223-236,ISSN 0022-1694 https://doi.org/10.1016/0022-1694(91)90139-9 (1991),

[23] Midhun M., Lekshmy P. R., Ramesh R., Yoshimura K., Sandeep, K. K., Kumar, S., et al. The effect of monsoon circulation on the stable isotopic composition of rainfall. Journal of Geophysical Research: Atmospheres, 123, 5205-5221. https://doi.org/10.1029/2017JD027427(2018)

[24] Gupta S. K. and Deshpande R. D. Synoptic hydrology of India from the data of isotopes in precipitation Current science, vol. 85, no. 11 (2003)

[25] Deshpande R. D., Maurya A. S., Kumar Bhishm, Sarkar A., Gupta S. K. Rain-vapor interaction and vapor source identification using stable isotopes from semiarid western India. Journal of Geophysical Research, vol. 115, d23311, https://doi.org/10.1029/2010JD014458, (2010)

[26] Sinha Nitesh, Chakraborty S., Mohan P.M., Modern rain-isotope data from Indian island and the mainland on the daily scale for the summer monsoon season, Data in Brief,Volume 23,2019,103793, ISSN 2352-3409 https://doi.org/10.1016/j.dib.2019.103793(2019)

[27] Unnikrishnan Warrier, C., Praveen Babu, M., Sudheesh, M., Deshpande, R. D. Studies on stable isotopic composition of daily rainfall from Kozhikode, Kerala, India. Isotopes in environmental and health studies, 52(3):219-230, https://doi.org/10.1080/10256016.2016.1103238 (2016).

[28] Chakraborty, S., Sinha, N., Chattopadhyay, R. et al. Atmospheric controls on the precipitation isotopes over the Andaman Islands, Bay of Bengal. Sci Rep6, 19555. https://doi.org/10.1038/srep19555 (2016)

[29] Munksgaard N.C., Kurita N., R. Murillo Sánchez et al. Data Descriptor: Daily observations of stable isotope ratios of rainfall in the tropics. Sci Rep9, 14419.https://doi.org/10.1038/s41598-019-50973-9 (2019).

[30] Sinha, N. and Chakraborty, S. Isotopic interaction and source moisture control on the isotopic composition of rainfall over the Bay of Bengal. Atmospheric Research. V. 35, 1 May 2020, 104760. https://doi.org/10.1016/j.atmosres.2019.104760 (2020)

[31] A.A Fousiya. 2017 Stable Isotopic studies of marine carbonates and precipitation: A tool to evaluate the surface ocean characteristics, Minicoy Island, Lakshadweep. Ph.D. Thesis, Anna Univeristy, Chennai, India, 166p; (unpublished). (2017)

[32] Chakraborty, S., Belekar, A.R., Datye, A. et al. Isotopic study of intraseasonal variations of plant transpiration: an alternative means to characterise the dry phases of monsoon. Sci Rep 8, 8647 (2018) https://doi.org/10.1038/s41598-018-26965-6 (2018)

[33] Bajaj, K., Thomas, R., Yadav, A., Datye, A., Chakraborty, S.,Hydrological Linkages between different Water Resources from two contrasting Ecosystems of western peninsular India: A stable isotope perspective. Isotope in Environment and Health Studies. https://doi.org/10.1080/10256016.2019.1666121 (2019)

[34] Lone, A.M., Achyuthan, H., Chakraborty, S., Metya, A., Datye, A., Kripalani, R.H.,Fousiya, A.A., Controls on the isotopic composition of daily precipitation characterized by dual moisture transport pathways at the monsoonal margin region of North-Western India, Journal of Hydrology, 588; 125106 https://doi.org/10.1016/j.jhydrol.2020.125106 (2020)

[35] Tanoue, M., Ichiyanagi, K., Yoshimura, K. et al. Seasonal variation in isotopic composition and the origin of precipitation over Bangladesh. Prog Earth Planet Sci5, 77. https://doi.org/10.1186/s40645-018-0231-4 (2018)

[36] Huffman, G. J., and Coauthors; The TRMM Multisatellite Precipitation Analysis (TMPA): Quasi-Global, Multiyear, Combined-Sensor Precipitation Estimates at Fine Scales. J. Hydrometeor., 8, 38-55, https://doi.org/10.1175/JHM560.1 (2007)

[37] Berkelhammer, M., Sinha, A., Mudelsee, M., Cheng, H., Yoshimura, K., and Biswas, J.: On the low-frequency component of the ENSO–Indian monsoon relationship: a paired proxy perspective, Clim. Past, 10, 733-744, https://doi.org/10.5194/cp-10-733-2014, (2014)

[38] Rajeevan M, Gadgil Sulochana, Bhate Jyoti : Active and break spells of the Indian summer monsoon J. Earth Syst. Sci.119, No. 3, June 2010, pp. 229-247, https://doi.org/10.1007/s12040-010-0019-4 (2010)

[39] Reference Sheet for VSMOW2 and SLAP2 international measurement standards. Issued 2009-02-13. International Atomic Energy Agency, Vienna. (available at https://nucleus.iaea.org/rpst/documents/VSMOW_SLAP.pdf) (2009)

[40] Dansgaard, W Stable isotopes in precipitation. Tellus, XVI: 436-468. (1964)

[41] Craig H Isotopic Variations in Meteoric Waters, Science,Vol 133, Issue DOI: 10.1126/science.133.3465.1702 (1961)

[42] Sarma D.D. Geostatistics with Applications in Earth Sciences, Springer, ISBN 978-1-4020-9380-7 (2009)

[43] Choudhury B., Saha S.K., Konwar M., Sujith K., & Deshamukhya A. Rapid drying of Northeast India in the last three decades: Climate change or natural variability? Journal of Geophysical Research: Atmospheres, 124, 227-237. https://doi.org/10.1029/2018JD029625 (2019)

[44] Sebastian F.M. Breitenbach, Jess F. Adkins, Hanno Meyer, Norbert Marwan, Kanikicharla Krishna Kumar, Gerald H. Haug,Strong influence of water vapor source dynamics on stable isotopes in precipitation observed in Southern Meghalaya, NE India,Earth and Planetary Science Letters,Volume 292, Issues 1-2,2010,https://doi.org/10.1016/j.epsl.2010.01.038. (2010)

[45] Risi, C., S. Bony, and F. Vimeux. Influence of convective processes on the isotopic composition (d18O and dD) of precipitation and water vapor in the tropics: 2. Physical interpretation of the amount effect, J. Geophys. Res., 113, D19306, https://doi.org/10.1029/2008JD009943. (2008)

[46] Ahmed, N., Kurita, N., Chowdhury, M. et al. Atmospheric factors controlling stable isotope variations in modern precipitation of the tropical region of Bangladesh. Isotopes in Environmental and Health Studies, 56, 220-237.

[47] Berkelhammer, M., Sinha, A., Stott, L., Cheng, H., Pausata, F.S.R., Yoshimura, K, 2012. An abrupt shift in the Indian monsoon 4000 years ago. Geophysical Monograph Series 198, 75-87, https://doi.org/10.1029/2012GM001207 (2012)

[48] Sinha, N., Gandhi, N., Chakraborty, S., Krishnan, R., Yadava, M.G., Ramesh, R. 2018. Abrupt climate change at ~ 2800 Yr B.P. evidenced by high-resolution oxygen isotopic record of a Stalagmite from Peninsular India. The Holocene 28 (11): 1720-1730 https://doi.org/10.1177/0959683618788647 (2019)

[49] Kathayat Gayatri, Cheng Hai, Sinha Ashish, Yi Liang, Li Xianglei, Zhang Haiwei, Li Hangying, Ning Youfeng, Edwards R. Lawrence. The Indian monsoon variability and civilization changes in the Indian subcontinent, Sci. Adv. 1-8 DOI: 10.1126/sciadv.1701296 (2017)

[50] Band, S and Yadava, M 2020 Addressing past monsoon variability from Speleothems. Curr. Sci., 119, 244-254